Small Wind Turbines for Electricity and Irrigation

Small Wind Turbines for Electricity and Irrigation
Design and Construction

Mario Alejandro Rosato

CRC Press
Taylor & Francis Group
Boca Raton London New York

CRC Press is an imprint of the
Taylor & Francis Group, an **Informa** business

CRC Press
Taylor & Francis Group
6000 Broken Sound Parkway NW, Suite 300
Boca Raton, FL 33487-2742

Contents

Foreword

I got acquainted with Mario during the research project *H2Ocean** (2012–2014, funded by the 7th Frame Program of the European Union), when I had the pleasure of collaborating with him and observing by myself his creative streak. When we met during plenary meetings of the cited project he often told me, while having breakfast: "Maurizio, last night I did not sleep well last night, I had another creativity attack…"

Mario has multifaceted competences, matured along years of experience in many fields, a proof of his curiosity and Renaissance intellect. In particular, I had the chance to discuss and contrast some ideas about wind turbines with him, one of the research fields we share and the one to which this book is dedicated.

One of his strong points, rare in people with a creative streak, is being able to develop an idea from the draft to the practical construction of the proto-type, its improvement, and the development up to its commercial stage. He is then able to conjugate the necessary abstraction and creativity to develop new ideas with pragmatism and, on top of all, tenacity, a quality perhaps more important than all the others. As T. A. Edison said: *"Genius is 1% inspi-ration, 99% perspiration."*

This book was written with such spirit: keeping the theory to the necessary minimum, it concentrates on the practical aspects; it is suitable for anybody who has interest in designing and building wind turbines for domestic usage, without losing sight of economic sustainability.

Good reading to all (of us).

24th August 2014
Dr. Maurizio Collu
Lecturer, Cranfield University
Cranfield, UK

---

* www.h2ocean-project.eu/.

Preface

> It is the tension between creativity and skepticism that has produced the stunning and unexpected findings of science.
>
> **Carl Sagan**

A quarter of a century has already passed since the publication in Spain of my first book, *Diseño de Máquinas Eólicas de Pequeña Potencia*. In those days, I had already reached the firm belief that small wind turbines, together with anaerobic digesters and thermal solar systems, are the most democratic of renewable energies, because anybody can build a windmill, a digester, or a thermal solar collector, even with scrap material, without depending on sophisticated technologies and exotic materials. Among the said three energy sources, small wind power installations provide mechanical work, which—in thermodynamic terms—is the highest quality of energy. Small wind power systems are then one of the three pillars for a sustainable society, based on circular economy.

At the end of the 1980s, in spite of the Chernobyl disaster and the growing perception of the negative environmental impact of fossil fuels, many people in the so-called industrialized societies still considered small stand-alone energy systems as suitable only for remote rural areas, where standard financial analysis shows that it does not pay off to bring an electric line. During the last decades, politicians, industrialists, and economists in all countries have often criticized and even opposed small-scale initiatives for the energy independence of families or communities. The general dogma of "scale economy" pushes in the direction of gigantic infrastructures that tend to perpetuate the linear business model, based on centralized generation (more or less monopolized), energy transmission (often along huge distances), and passive, energy-dependent consumers. Society has evolved, and a growing number of citizens became aware that modern society needs to change its consumerist attitude; otherwise the planet will not survive. Large sectors of the population now search for their energy independence, based on renewable sources, even when traditional—reductive—economic criteria seem to show that taking electricity and gas from the grid is cheaper. In some countries, for instance, Spain, the political class has virtually banned the installation of small stand-alone energy systems by taxing and putting unnecessary bureaucratic burden on them, in an attempt to protect the interests of big energy companies. In the time elapsed between the publication of my first book and the present, the population has grown from 5.5 billion to 7.5 billion souls. Climate change has become more severe and the number of people lacking access to energy has grown exponentially, both in developing and industrialized countries. Increasingly, severe draughts and desertification are the cause of widespread social tensions and war, which push entire populations to abandon their homelands, seeking a better life. Yet the solution to such global problems could be very simple and does not require high investments: empowering people to adopt circular economy practices and promote food, water, and energy autonomy. There is no need for high-tech solutions, nor is there a universal technology that can work efficiently at any scale and in any geographical context. Small wind power systems are just one of the alternatives in the designer's or the decision maker's toolbox to provide a simple solution to mankind's growing demand for energy.

With much of Carl Sagan's creative and skeptic spirit, and the retrospective vision of almost three decades of evolution of the renewable energies market, I decided to write this book. This time in English, in order to reach a wider public.

One may argue that there are plenty of manufacturers of cheap and nice-looking small wind turbines, with powers ranging from fractions of kW to tens of kW. What then is the use of a book on small wind turbines? The answer is simply that in the last three decades small wind turbines have evolved in the shadow of big ones, adopting design criteria inherited from the engineering of large wind farms. The latter are usually installed in places accurately chosen because of their high wind potential. On the other hand, single users needing a small wind turbine to pump water for agriculture or to provide the basic electricity supply to their homes cannot choose the site with highest average wind speed and the highest capacity factor. Commercially available small wind turbines are then unsuitable to solve the problem they are expected to. In the water pumping segment, windmills are still anchored to 19th-century designs, having low aerodynamic efficiency and high material cost.

The Author has conceived this book as a practical manual for:

1. Undergraduate students interested in renewable energies.

2. Engineers who need to design an energy autonomous home for a customer, not necessarily engineering and building themselves the turbine, but requiring a rational method on how to choose the most suitable model among a myriad of available products.

3. NGO agents and collaborators, who need to provide a punctual energy supply during humanitarian disasters or for improving poor people's life within long-term development projects.

4. Citizens willing to achieve energy self-sufficiency and independence from fossil fuels, either for ideological reasons or for pure need.

5. Farmers needing to cope with the increasing cost of conventional water pumping for their crops and animals.

6. Small industries needing to reduce their energy bills in order to strive in a global competitive market.

7. Politicians and decision makers, even those having no technical background, willing to understand why small autonomous energy systems in general, and small wind turbines in particular, are key for reaching the paradigm of a sustainable society.

8. Business Angels or investors, who are often offered with "disruptive inventions," will find here all the background knowledge to understand why all such technologies are always condemned to fail.

9. Public officers, who sometimes face the request of funding research projects on "emerging technologies," will find here the elements to judge why a 3-bladed horizontal wind turbine has not been beaten yet by any "disruptive invention." Not wasting contributor's money in dead-end research is as important as funding true research with market potential.

10. Small entrepreneurs or workshops wishing to start a new brand, or to diversify their current production, bringing specifically designed products to their local markets.

11. Managers of energy utilities—although it may seem a paradox—who from time to time need to solve the problem of "serving a load, but an electric line does not pay off."

Each chapter of this book deliberately follows the same didactic scheme: a theoretical explanation, limited to the minimum but as rigorous as possible, and some practical examples showing the calculation procedures. A set of simple spreadsheets is available for downloading, each providing further examples on how to solve some specific design problem, and allowing the reader to "play" by changing parameters and seeing "what if..." This simple trial and error learning process allows the beginner to develop the feeling of the orders of magnitude involved in the design of a small wind power system, its potential advantages on other alternative solutions—as photovoltaic panels or diesel generators—and eventually its limitations under some special circumstances.

This book covers the most common wind turbine technologies in the small power segment: horizontal axis, both for electrical generation and water pumping, vertical axis of the Darrieus type, and vertical axis of the Savonius type. Chapter 11 is a collection of "alternative" or "unconventional" wind power technologies that never reached maturity, with short explanations about their logical flaws and practical limitations. Finally, Chapter 12 is a compendium of tables, formulas, and aerodynamic data on airfoils, streamlined and blunt bodies, especially at low Reynolds numbers, which is the most common situation in the small power segment.

I want to thank Taylor & Francis Group for having the courage and open mindedness to endeavor the publication of a book that is somehow unconventional in its approach, and in some ways questions the established criteria for the design and operation of renewable energy systems, and the dogmas of scale economy.

Mario Alejandro Rosato
Latisana (Italy), 29th January 2018

Author

Mario Alejandro Rosato, electric-electronic and environmental engineer, graduated in 1988 with a thesis on the construction of a low-cost, small-sized wind turbine, meant to supply electricity to rural schools in the Argentine Patagonia. In 1992, the first version of this book was published in Spain, which was focused on self-construction of wind turbines in developing countries. In 2003, he won the *John Hogg Award* with a simple program, based on a spreadsheet, to the design of wind turbines meant for alternative marine propulsion. In 2010, the renewed interest of the Italian public opinion on small-sized wind turbines led him to start teaching the subject at the University Consortium of Pordenone and other private institutes. In 2012, he was engaged by Acca Software, the Italian leader in software and technical training for engineers, to record a video course on how to select commercial wind turbines and design the whole system. In 2014, with the help of his wife Giovanna he decided to translate the original book into Italian but, given the technological evolution in nearly one quarter of century, he finally decided to write a completely new book from a scratch.

The current edition is a not a mere translation to English of the Italian one. The Author included an additional chapter and further improvements of the pre-existing ones, and has adapted the content to meet the different needs of people living both in industrialized and in developing countries. The formulas and design methods are presented together with some spreadsheets that will allow the reader to design simple turbines and installations, without the need to resort to expensive simulation software. Each chapter includes a few solved examples.

1

Small Wind Turbines: A Technology for Energy Independence and Sustainable Agriculture

1.1 Introduction: Why "Small" Wind Turbines?

This book is an improved English version of the 2015 edition in Italian, and a radical review and update of the first studies carried out by the Author in Argentina, published in Spain in 1992. The Author's scope is to fill some gaps in the literature about wind turbines, since most of it addresses the design challenges of big machines (i.e., bigger than 16m in diameter) and the economical optimization of onshore or offshore wind farms. There is a wide array of specific texts on small wind turbines available, but they are mainly do-it-yourself manuals, just valid for assembling a given model, definitely not universally applicable to any site and user requirements. Other books are academic treatises on aerodynamics at low Reynolds numbers and numerical simulation of unsteady conditions, not suitable for those readers lacking the specific knowledge on computer-aided fluid dynamics (CAFD), or the economic means to access sophisticated calculation models. In all cases, the focus of the literature is on turbines for electric generation. The reason the Author decided to endeavor to write a book on small wind turbines, including both horizontal and vertical axis types, both action and reaction families, and both for electricity and for water pumping, is because small wind turbines are a key factor for creating a decarbonized economy and ensuring sustainable development for future generations. The reasoning supporting the Author's conviction is both ideological and technological. From an ideological point of view, the *Universal Declaration of Human Rights* has a flaw: it does not expressly declare the right to energy self-sufficiency and the right to a minimum energy availability that ensures a decent level of life. One could argue that the right to energy self-sufficiency derives from Articles 3, 17 (1), and 25 (1):

Article 3.

Everyone has the right to life, **liberty** and security of person.

Article 17.

(1) Everyone has the right to **own property** alone as well as in association with others.

Article 25.

(1) Everyone has the right to a **standard of living adequate for the health and well-being of himself and of his family**, including food, clothing, housing and medical care and necessary social services, and the right to security in the event of unemployment, sickness, disability, widowhood, old age or other lack of livelihood in circumstances beyond his control.

Hence, any individual should have the **liberty** (intended as freedom of choice) to self-produce energy and the **right to own** it, i.e., being free either to consume it or to sell it to anybody, without the obligation of including private or public intermediary traders in the transaction. By extension, everybody should have the liberty to choose the most suitable technology, or combination of technologies, to grant him/her an **adequate standard of living**.

The availability of energy allows replacing human physical work with machine work, especially for very heavy or tedious tasks. Consequently, having access to energy means having more free time, both for education and self-realization. Free time and education are two human rights, expressly listed by Articles 24 and 26 (1):

Article 24.

Everyone has the right to rest and leisure, including reasonable limitation of working hours and periodic holidays with pay.

Article 26.

(1) Everyone has the right to education.

On the other hand, the words "energy" and "water" are not expressly mentioned anywhere in the Declaration, and Article 25 does not define what an "adequate" standard of living is. The United Nations (UN) General Assembly officially recognized the **Human Right to Water and Sanitation (HRWS)** on July 28, 2010. Furthermore, the World Health Organization (WHO) has defined that the minimum basic requirement of clean water for drinking and basic hygiene is 20 l per capita/day. As of May 2017, neither the UN, nor any individual State, have expressly recognized the human right to energy, no "minimum vital energy availability" has been defined, and the right to energy independence is still a matter of juridical and academic debate, especially in industrialized countries. A simple Internet search with the keywords "human right to energy" yields more than 132 million pages, the first 50 results being mainly university research articles, theses, and pages of environmentalist associations. The bibliography at the end of this chapter cites two articles found in the UN's official website for those readers wanting further information.

The missing official recognition of energy self-sufficiency as a human right has led to the promulgation of some controversial laws in some countries that favor big utility companies and hinder private citizens and small companies wishing to become energy-independent. The following two examples are the ones better known to the Author, though it is probable that similar situations exist in many other countries around the world:

a. Spain's ill-famed "Sunlight Tax."

The Royal Decree 900/2015 virtually forbids the installation of stand-alone electric generators and accumulators for self-consumption. All electricity generators must be connected to the national grid and a tax must be paid for it. The direct trade of self-produced energy (i.e., producing energy in a home and placing a cable to sell energy to the neighbor) is forbidden. It is called "Sunlight Tax" because most small power energy plants in Spain installed before 2015 are photovoltaic. Furthermore, the Government has eliminated the dispatching priority for renewable energies, which means, even when there is plenty of sun and wind, the utility companies can opt for not allowing the injection of such energy in the grid, and satisfy the demand with their nuclear, gas- and coal-powered plants.

b. Italy's proverbial bureaucracy.

Italian laws subsidize the generation of electricity from renewable sources. Such support is more formal than practical, because the policy of subsidies is strongly biased either in favor of some technologies or of certain plant sizes. Photovoltaic electricity used to be by far more subsidized than wind power and the procedure to obtain the necessary permits is generally quicker for photovoltaic panels than for wind turbines. A certain percentage of renewable energy is mandatory in planning the construction of new buildings, but the only technologies considered are photovoltaic, solar thermal, geothermal, and biomass. In spite of their low environmental impact compared to big wind farms, small-sized wind turbines are rare in Italy, because local authorities require the same amount of paperwork, regardless of the size of the installation. A private citizen, or a small company, intending to install a small wind turbine for self-production will be requested to perform a three-year anemometric campaign to demonstrate that the site is suitable, a study of the visual and environmental impact, a study on seismic and hydrogeological risk, structural tests, and certification of every single component, etc. The large amount of paperwork and the long delay to obtain the permits discourage most of the potential users. As in Spain, all the energy produced must be exchanged with the national grid and it is not legal for a citizen or a company to sell electricity directly to another consumer.

We will leave to experts in law and international High Courts the task of deciding whether the right to energy self-sufficiency must be considered as a human right, and if national laws like the "Sunshine Tax" are violating such right. The second part of our reasoning will analyze, from the sociological and physics points of view, why the free access to energy is so important to fulfill Articles 24, 25, and 26 of the Declaration of Human Rights. Furthermore, we will try to determine in a scientific manner how much energy is necessary to ensure a minimum standard of living that can be considered "adequate" in compliance with Article 25 of the Human Rights Declaration. Finally, we will analyze why small wind turbines are more suitable than other technologies to reach all of the above goals.

The International Energy Agency (IEA) has a web page dedicated to the topic of energy poverty, but does not provide an explicit definition of it. From the corresponding webpage, https://www.iea.org/topics/energypoverty/, an implicit definition of energy poverty is the lack of

> ...access to modern energy services. These services are defined as household access to electricity and clean cooking facilities (e.g. fuels and stoves that do not cause air pollution in houses).

Reading the cited webpage, one could deduce that the IEA addresses energy poverty exclusively from the point of view of developing countries in Asia and Africa, and that their only concern is the smoke in the houses. Further reading better clarifies the point:

> Modern energy services are crucial to human well-being and to a country's economic development; and yet globally 1.2 billion people are without access to electricity and more than 2.7 billion people are without clean cooking facilities. More than 95% of these people are either in sub-Saharan African or developing Asia, and around 80% are in rural areas.

In any case, the IEA does not quantify a threshold to define when an individual or household is "energy poor."

The European Community has not issued an official position on energy poverty, but a page of its web site contains a study on the topic, available to download (see Bibliography at the end of this chapter). The said document defines energy poverty as:

> Energy poverty, often defined as a situation where individuals or households are not able to adequately heat or provide other required energy services in their homes at affordable cost.

Such definition is, in the Author's point of view, somehow restrictive and addresses the problem only from the perspective of Central and Northern European countries, where the main energy consumption is heating in winter.

In the UK, energy poverty is restrictively called "fuel poverty." It is defined and quantified—though in an arbitrary manner—as: "needing more than 10% of the household's income on fuel to maintain an adequate level of warmth."

Another unofficial document on energy poverty in Europe is the *Handbook of energy poverty*, prepared and published by the office of Tamás Meszerics (Member of the European Parliament) and reported in this chapter's Bibliography. The same contains a broader definition of energy poverty, though admitting that it is very difficult to quantify its threshold.

> Energy Poverty is commonly understood to be when a person or household is not able to heat or fuel their home to an acceptable standard at an affordable cost. In reality, it covers a very wide set of essential activities. It can occur if people cannot afford to heat their homes adequately, but also to cool them in hot climates. It may mean they cannot afford to cook hot meals, or have reliable hot water for baths and washing clothes or run essential domestic appliances (washing machines, irons, televisions, computers, etc.).

In the United States, the concept of energy poverty is encompassed in the broader definition of poverty. The Government's approach is just defining thresholds of poverty, based purely on the individual's or household's income and its geographical residence, and applying a tabulated subsidies policy.

The common denominator of all the approaches to energy poverty cited so far is that none of them recognizes the right to energy independence. The approach in the European Union (EU) and United States, on the contrary, creates more dependency from the utilities by giving subsidies to pay for the bills, which may be useful for a transitory contingence, but provides no solution to the problem in the long term.

Popular wisdom says that, if you want to help a hungry person, you should not give him/her a fish to eat, because tomorrow he/she will be hungry again. Instead, give him/her a fishing rod and the basic instruction on how to fish. With such spirit, let us try to figure out "how big the fishing rod must be," i.e., how much energy does a family require to reach a *"standard of living adequate for the health and well-being,"* as stated by Article 25 of the Declaration of Human Rights. It is clear that the energy demand will depend strongly on the climate, the number of individuals in the household, and their age and health state. This is probably the reason for the discrepancies about the definition of energy poverty found so far. Hence, we will split the problem in two:

a. Baseline energy consumption. This is the quantity of energy universally necessary for ensuring a person's self-realization, and physical and psychological wellness with state-of-the-art technology. It is independent of climatic, geographical, or social factors.

b. Climate-dependent demand (heating/air conditioning) and local contingencies or lifestyle needs (e.g., daily mobility requirements). This amount of energy is additional to the one formerly defined, but local policy makers should calculate it case by case.

The concept underlying the Author's next reasoning is that motors and other energy transformers replace people in heavy, tedious, dangerous, or unhealthy work, freeing time in the daily life of an individual for self-realization. In the same way, computers, radios, television sets, and mobile phones are all means allowing access to information, and in broad sense, they enable the right to education.

It is probable that the recognition of the right to a minimum amount of water as a human right derives from the fact that anybody can understand how much water a person needs to drink, cook food, and have a shower. On the other hand, maybe the right to a minimum energy availability has not been recognized as a human right so far because it is hard to figure out what the availability of 1 kWh of electricity means. In order to demonstrate how important energy availability is, and why it must be considered a human right, we will develop here the concept of "virtual slave."

According to Avallone (2007), a healthy, well-fed laborer over the course of an eight-hour work shift can sustain an average output equal to 0.075 kW, i.e., the energy produced by such a worker during a work shift is 0.6 kWh.

With such equivalence in mind, we will calculate how many "virtual slaves" are necessary to ensure a minimum standard of life to a family of four individuals. The water needs are: 20 l per capita/day for drinking, food preparation, and cooking, plus 10 l/day for personal hygiene, plus 10 l/day for laundry washing, plus 10 l/day for cleaning home, plus 10 l/day for growing basic food, plus 10 l/day for sanitation and waste disposal. In total 70 l/day per capita, i.e., 280 l/day. Said values derive from a publication of the WHO (see Bibliography). Please note that such standard is a minimum conservative calculation; the average per capita water consumption in the EU is 102 l/day (official statistics www.eea.europa.eu/data-and-maps/indicators/use-of-freshwater-resources-2/assessment-2), though 50 l per capita/day may be enough in urban contexts and with a responsible use of the resource. We will assume conventionally that the water must be pumped from 30 m depth, in order to assign it a minimum value of incorporated energy, though safe drinking water usually requires pumping from deeper wells. The basic household appliances required for food preparation and conservation, entertainment and information or education, lighting, and hygiene are:

- A cook stove, total energy consumption 3 kW × 2 h/day.
- A vertical refrigerator, 132 l capacity, total power 90 W, average consumption 0.75 kWh/day.
- A washing machine, 0.5 kWh daily consumption (yearly average).

TABLE 1.1

Baseline Energy Demand for a "Standard of Living Adequate for the Health and Well-Being," in the Case of a Generic Four-Member Family

Activity	Energy Demand (kWh/day)	Virtual Slaves
Cooking food	6.0	10.0
Conserving food (cooling)	0.8	1.3
Washing clothes	0.5	0.8
Communications/Entertainment	0.3	0.5
Lighting	1.5	2.5
Water supply	0.03	0.1
Sanitary water for basic hygiene	1.0	1.7
Total	10.1	16.8
Total per capita	2.5	4.2

- A generic device for communications and entertainment (TV, PC, or radio transponder). Assumed as $0.10\,\mathrm{kW} \times 3\,\mathrm{h/day}$.
- Lighting, assumed with conventional halogen lamps $5 \times 0.075\,\mathrm{kW} \times 4\,\mathrm{h/day}$.

The energy required for the water supply, assuming 60% overall efficiency in water pumping, is:

$$E_{water} = \frac{280\,\mathrm{kg/day} \times 20\,\mathrm{m} \times 9.8\,\mathrm{m/s^2}}{0.6} = 91.5\,\mathrm{kJ/day} = 0.025\,\mathrm{kWh/day}$$

The energy required for essential sanitary hot water, assuming 20°C of temperature increase and 90% heating efficiency is:

$$E_{sanitary} = \frac{40\,\mathrm{kg/day} \times 20°C \times 4.184\,\mathrm{kJ/kg°C}}{0.9} = 1\,\mathrm{kWh/day}$$

The baseline per capita energy demand for a household as the one assumed, and the resulting equivalent "virtual slaves," are summarized in Table 1.1.

Rounding values, we can state with good accuracy that an individual must have access to at least 2.5 kWh/day in order to satisfy his/her needs for a minimum standard of living. Failing to grant such minimum energy input to the population has the following consequences:

a. diseases arising out of lack of hygiene and improper food preparation or conservation;

b. people forced to walk long distances to procure water and firewood (in developing countries);

c. people forced to beg for charity in the street or wait in the line to ask for subsidies to pay for their bills, which leads to growing indebtedness, and in turn, may lead to popular protests against utility companies, banks, and social tensions in general (in industrialized countries);

d. consuming time in such activities like b) or c) subtracts time from education or, in broader terms, to access to information. Consequences of increasing levels of ignorance include: a weaker awareness and exercise of civil rights, religious fanaticism and intolerance, and a sense of frustration and social class hatred that opens the doors for extreme populist and nationalist groups to take profit of such situations.

By now, it should be clear why the right to a minimum availability of energy, and the right to choose freely whether to self-produce such energy, deserves an explicit recognition as a human right. Furthermore, we have quantified on a rational basis the minimum energy availability necessary to grant a decent standard of living, regardless of geographic (climatic) factors, which must be accounted on top of the baseline, considering the local needs for heating, mobility, etc. Finally, the concept of "virtual slave" gives a tangible meaning to the more abstract unit "kWh." In other words, if energy is not available, people must replace it with man power, becoming then "real slaves" for mere survival. Figures 1.1 and 1.2 show very eloquently the worst consequences of energy poverty in developing countries: people, usually girls and women, forced to consume their lives procuring firewood and water, just to satisfy the most elementary survival needs of their families. Such hard, time-consuming, physical labor precludes them from any further possibility of progress and self-realization.

FIGURE 1.1
Photo by TAPUWA MASAWI published in *News from the South*, http://newsofthesouth.com/energy-crisis-in-rural-areas-women-bear-the-brunt/, republished with permission.

FIGURE 1.2
Mrs. Jangali Ram pumping water for her family. Photo by Jim Holmes for AusAID. (13/2529), Nepal, 14th June 2013, republished under Creative Commons license BY02.

Now it is time to answer the question at the beginning of this section: *"Why small wind turbines?"* The answer is just a consequence of the Laws of Thermodynamics. Wind is mechanical energy, which can be converted with high efficiency in practically any other form of energy, and if necessary, it can be stored too. Wind turbines are then the perfect "virtual slaves" to relieve people from heavy, time-consuming work. Having a look at Table 1.1, it appears evident that a decent living standard requires not only 2.5 kWh/day per person, but also that such amount of energy is required as heat or as electricity, or as pure mechanical work, depending on the need. A single wind turbine producing electricity is then able to satisfy any of the basic energy requirements of a household identified in Table 1.1, because mechanical and electrical energies are easily convertible in heat or potential energy, and can be stored in many ways. One could argue that a photovoltaic panel could produce the same amount of electricity as a wind turbine if correctly dimensioned. Fully true, but there are strong differences between both technologies when it comes to their sustainability:

a. Wind turbines are effective at any latitude, even at the Poles or on top of high mountains (especially on the latter!), while photovoltaic panels are progressively more ineffective at latitudes higher than 45° and snow at high quotas reduces their yield.

b. Photovoltaic panels lose efficiency with time, roughly 1% every year. Wind turbines will not lose their efficiency with time or, in the worst conditions, they may lose a very small fraction of their efficiency because of the additional rugosity caused by dust carried by the wind, but such efficiency loss will remain constant in time (this will be explained in Section 4.2.3).

c. Wind turbines have long service lives, and are easy to dismantle and recycle at the end. In contrast, dismissed photovoltaic panels are classified as e-waste (electronic waste) and are potentially noxious to the environment because they contain rare-earth elements. Their recyclability is arguable (potentially toxic fumes emitted from the plastic parts welded to glass and silicon chips when separating them), and the amount of energy necessary to recover valuable rare elements is high.

d. Anybody can build, maintain, and repair a small wind turbine by himself, with little initial capital and, in some cases, employing scrap materials and components. On the other hand, solar panels are high-technology items that only a few companies in the world are able to produce. Expecting to become energy-independent with the installation of a photovoltaic system is an illusion: the owner may become independent from the utility company, but in turn will become technology-dependent on the system's manufacturer for the spare parts. Installing a photovoltaic system requires an investment that only upper-middle class people in industrialized countries can afford.

e. Building and installing small wind turbines where energy-poor people need them can prove to be a way to create small industries and jobs locally. On the contrary, importing photovoltaic panels just make multinationals become richer, poor people become poorer, and the middle class pays for the subsidies' bill.

1.2 Why Not "Big" Wind Turbines?

The general tendency throughout the world has been a kind of competition between multinational companies to build giant wind turbines—each model bigger than the former—and to install them in large wind farms. The usual argument is "economy scale," though most probably it is a strategy for establishing technological barriers to the entry of smaller competitors. As of May 2017, the biggest wind turbine in the world is manufactured by the Danish multinational Vestas, in cooperation with Mitsubishi Heavy Industries. Its diameter is 164 m and has 8 MW rated power.

Such approach is unsustainable in the long term, because of several reasons:

a. Social imbalances
   i. Big wind farms perpetuate the old model of centralized energy production, transmission, and distribution (usually a monopoly or oligopoly) to consumers, who are absolutely passive in the process.

ii. Wind power is "non-dispatchable" energy, i.e., it is difficult to program its offer and to adapt it to the instantaneous energy demand. Energy accumulation on a large, centralized scale, together with keeping the stability of the grid's frequency, are big technological challenges and the cost of managing and/or accumulating a random production of energy is high. Often Governments have imposed to utility companies the obligation to include given percentages of renewable energy in the generation mix. Utility companies have then charged all the inefficiencies arising from managing an obsolete grid to the energy consumers, instead of investing in converting their infrastructure in smart grids. This has created the perception among consumers that renewable energies are the cause of increased energy prices, and hence of energy poverty. All said problems could be partly resolved by reducing or eliminating the bureaucratic procedures to install and run small wind turbines, and eliminating the obligation to connect them to the grid in those countries where such rule applies.

iii. It happens sometimes that, in order to obtain the necessary permits to install a wind farm in a site with much potential, big multinationals corrupt local administrators, or resort to "generalized compensations" to those living near to the chosen site, for the "acknowledged impact" of the installation. The cost of corruption, or of "compensations," always ends up increasing the final price of the energy paid by the rest of the society.

b. Environmental impact

Big wind farms, either onshore or offshore, pose environmental challenges. Beyond the evident visual impact on the landscape, large wind farms can disturb the fauna and compromise the reproductive season. Dr. Manuela de Lucas, researcher of the National Park of Doñana, Spain, has been studying the effect of a big wind farm on the local birds since 2003. Her studies have demonstrated that the population of the European vulture (*Gyps fulvus*), an endangered species, has decreased quickly since the construction of the wind farm near the borders of the National Park. Nearly 5,000 dead birds have been counted at the feet of the turbines every year, of which 111 belong to the rare vulture species. The same researcher has encountered a proportional increase in the population of foxes and other small carnivores, who eat the carcasses of the dead birds. According to an article in *American Scientist*, June 2016, an increasing number of dead bats has been observed near wind farms in the United States and Canada. Politicians and utility companies have tried to perpetuate the model of centralized, monopolistic energy generation, and at the same time avoid the so-called NIMBY syndrome (not in my back yard) of local populations, by placing wind farms offshore. Some people argue that the environmental impact of such systems is favorable, because the no-fishing area becomes a kind of sanctuary where fish and mollusks can reproduce. Nevertheless, it is not clear what effect the colossal foundations of such huge turbines have on the *benthos*, the sea-bottom ecosystem.

c. Effect on local economies

There are very few constructors of large wind turbines in the world; usually having their manufacture facilities far away from the places where the turbines are installed. For sure, such manufacture facilities create jobs in the manufacturer's country and profit for the investors owning the wind farm, but the community living in the area where the turbines are installed usually receives no benefit. Areas living on tourism can even suffer a decline in the number of visitors because of the alteration of the landscape. From this point of view, the installation in a given territory of 1,000 turbines of 1 kW power each would create a fairer share of the benefits than the installation of a single 1 MW turbine; moreover, with less impact on the landscape and fauna (Figure 1.3).

FIGURE 1.3
Near-shore wind farm of Vindeby. This was the first offshore wind farm in the world, it is located near Lolland, a small town in Denmark (Photo by the Author, 2009). According to local people, eels and other fish were disappearing because of commercial overfishing, but slowly began to recover because the no-sailing, no fishing area occupied by the wind farm has turned to be a kind of sanctuary where fish can reproduce and grow. Positive environmental impact? As of December 2017, the wind farm is under decommissioning.

1.3 How Small Are Hence, "Small" Wind Turbines?

We will demonstrate in Chapter 2 that, for a given wind speed, the power of a wind turbine is proportional to the square of its diameter. Figure 1.4 illustrates the concept.

Conversely, for a given diameter of the rotor, the output power grows with the cube of the wind speed. It is necessary to remark that people need energy, not power, to satisfy their needs. This concept is important, since many manufacturers of small wind turbines usually offer models having relatively high-rated power, but with relatively small diameters (cheaper rotor). The trick is that, in doing so, the manufacturer makes his models appear economically more convenient, which easily convinces potential customers. Turbines designed in such way will seldom meet the user's energy demand, unless the site is very windy. Since energy is the product of the power by time, the size of a wind turbine to meet a given energy demand will be a function of the wind speed distribution, i.e., the number of hours that the wind blows in the design speed interval.

According to the international norm, **IEC 61400-2:2013 Wind turbines - Part 2: Small wind turbines**, a turbine is considered "small" when the area swept by its rotor is smaller than 200 m², i.e., when $D < 16$ m. In "normal" contexts (class III and IV, low and very low wind locations) this means less than 20 kW rated power.

1.4 Why Small Wind Turbines for Pumping Water?

Earth's climate is becoming warmer and drier, and at the same time, the population continues to grow. Vast areas are becoming arid and unproductive, forcing local populations to migrate in search of better life conditions.

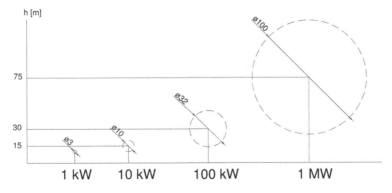

FIGURE 1.4
Size comparison of horizontal axis wind turbines at the reference speed of 8 m/s.

In order to mitigate such dramatic effects, and to produce food for a growing population, it will be necessary to recover the productivity of abandoned land, which means extracting water from underground at increasing depth, which in turn requires huge amounts of energy. Wind-driven water pumps are the ideal technology to reach such scope: they are simple and easy to manufacture and maintain, do not require the construction of an energy distribution infrastructure, and moreover, storing water is easier and cheaper than storing electricity. A few figures will help foster appreciation for the role that small wind turbines can play in the near future. Imagine having to cultivate horticultural crops in a sub-Saharan context. Assuming a good management of the culture with modern techniques, it will be necessary to supply roughly 4,000 m³/ha during the dry season (June–October), i.e., 22 m³/day. Supposing that the average well depth is 75 m, and the overall efficiency of the pumping system is 70%, the amount of primary energy to extract the water necessary for the crop will be:

$$E_{pumping} = \frac{22,000\,\text{kg/day} \times 9.8\,\text{m/s}^2 \times 75\,\text{m}}{0.7} = 6.41\,\text{kWh/day}$$

Pumping water with a petrol-driven pump would require from 1.5 to 2 kg of fuel per day per each irrigated hectare (1.7–2.3 l/day), plus the maintenance costs of the motor. A wind turbine directly coupled to a pump, running 24 h/day at an average 0.3 kW power, will do the job without noise, nor contamination, nor dependency from the external supply of fuel, lubricating oil, maintenance service, and spare parts.

It may seem incredible, but windmills for pumping water have remained almost unchanged since the conquest of the *Far West*, and hence the current offer of commercial windmills is obsolete, leaving a wide margin for small entrepreneurs to conquer a market that is destined to grow in the near future.

1.5 General Plan of This Book and Acknowledgments

The structure of this book has been thought to facilitate its reading and practical application, even for readers with only basic notions of Mathematics. Each chapter contains a section with the basic theory, a section with practical applications and a section with exercises and numerical examples. The following lines summarize the content of the book:

- Chapter 1 explains the Author's motivations to write a book specifically focused on small wind turbines and their use both for electricity generation and water pumping.
- Chapter 2 summarizes the basic theory of Aerodynamics. It is meant for undergraduate students and beginners in this fascinating

science, and is a useful review for engineers. This chapter includes the nomenclature and definitions that the Author has kept consistent throughout the subsequent chapters.

- Chapter 3 presents, in a reasoned manner and without omitting algebraic passages, a simplified, yet rigorous theory on the operation principles of horizontal axis rotors. A series of practical conclusions arising from such theory will be applied during the design of a horizontal axis rotor, explained in the successive chapter.

- Chapter 4 explains in detail, step by step, the design process of a horizontal axis wind turbine and its possible constructive variants.

- Chapter 5 presents a simplified theory on the design of *Darrieus*-type vertical axis wind turbines. Even though it is a simplified mathematical model, the results obtained with it are reasonably accurate for practical purposes. The second part of the chapter develops a practical example.

- Chapter 6 explains how to design and build *Savonius*-type vertical axis wind turbines. Said type of turbine is the most suitable for rural development, especially in low-income areas featuring weak winds.

- Chapter 7 contains the calculation method of the support structures of wind turbines, and includes some practical examples.

- Chapter 8 explains, without theorems and theories of statistical mathematics, how to practically employ the *Weibull* and *Rayleigh* functions for the estimation of the wind power potential of a site, when detailed anemometric data are not available.

- Chapter 9 explains in a practical manner how to select the most suitable battery type for each situation where a stand-alone system is required, how to interpret the technical data provided by battery manufacturers, and how to design an energy storage system.

- Chapter 10 explains how to dimension wind-powered pumping systems and provides some hints about alternatives to the classical single-effect piston pump.

- Chapter 11 contains a review of some unconventional wind-driven machines and the explanations on why they will never work; some examples of hypes in commercial catalogues and hoaxes on "super-Betz" technologies. The general information on old abandoned technologies periodically "rediscovered," and some examples on how to analyze "non-turbine" wind motors, will facilitate checking the information contained in catalogues of commercial wind turbines. Readers planning to invest in startups boasting an innovative wind power technology will find here a useful guide on failed or improbable solutions, which can eventually save them from frauds.

- Chapter 12 is a compendium of tables of aerodynamic data of blunt and streamlined bodies (profiles and nacelles) and a selection of airfoils, suitable for the design of aerodynamic action rotors. Although a large number of sources provide data on thousands of airfoils for aeronautics and aircraft models, the selection proposed in this chapter focuses on those profiles that are particularly suitable for the construction of wind turbines. The selection criteria were two: either because the recommended airfoil is easy to manufacture with simple tools, or because its aerodynamic performance is good with low Reynolds numbers and high surface rugosity.

The Author desires to thank prof. designer Mario Barbaro for the CAD elaboration of many of the drawings; arch. Giovanna Barbaro for her contributions and adaptations of many of the graphics; and prof. Maurizio Collu, University of Cranfield, for the critical review and valuable suggestions. Special thanks to prof. Trevor Letcher, Emeritus Professor Kwa-Zulu Natal University, South Africa, for having suggested the inclusion of a section at the end of each chapter, containing problems for students and their solutions.

Bibliography

Avallone E.A., Baumeister III T., and Sadegh A.M., *Marks' Standard Handbook for Mechanical Engineers*, 11th Edition, Mc-Graw Hill, New York, 2007, pages 9–14. ISBN 0-07-142867-4.

Bradbrook A.J., *Access to Energy Services in a Human Rights Framework*, Book proposal, file created on 23rd August 2005, saved in the UN site: http://www.un.org/esa/sustdev/sdissues/energy/op/parliamentarian_forum/bradbrook_hr.pdf.

Corbetta G., Pineda I., Azau S. et al., *Wind in Power-2013 European Statistics*, febbraio 2014. Pubblicazione della EWEA scaricabile gratuitamente da http://www.ewea.org/fileadmin/files/library/publications/statistics/EWEA_Annual_Statistics_2013.pdf.

IEA, *Key World Energy Statistics*, International Energy Agency (IEA), 9 rue de la Fédération, 75739 Paris Cedex 15, France, 2012. Scaricabile da https://www.iea.org/publications/freepublications/publication/kwes.pdf.

IEC 61400-2:2013, *Wind Turbines - Part 2: Small Wind Turbines*, International Standard, 3rd edition, 12th December 2013. https://webstore.iec.ch/publication/5433.

De Lucas M., Ferrer M., Bechard M.J., and Muñoz A.R., Griffon vulture mortality at wind farms in southern Spain: distribution of fatalities and active mitigation measures, *Biological Conservation* 147 (1), 184–189, 2012.

Mathews Amos, Amy, *Bat Killings by Wind Energy Turbines Continue - Industry Plan to Reduce Deadly Effects of Blades May Not Be Enough, Some Scientists Say*; American Scientist, June 7, 2016. Online version https://www.scientificamerican.com/article/bat-killings-by-wind-energy-turbines-continue/.

Steve Pye, Audrey Dobbins, *Energy Poverty and Vulnerable Consumers in the Energy Sector Across the EU: Analysis of Policies and Measures*. Insight_E Policy Report n. 2, May 2015 https://ec.europa.eu/energy/en/news/energy-poverty-may-affect-nearly-11-eu-population.

Tamás Meszerics, *Energy Poverty Handbook*, edited by the office of Tamás Meszerics (Member of the European Parliament) via The Greens/EFA group of the European Parliament. Brussels, October 2016, CAT: QA-06-16-183-EN-C, CAT: QA-06-16-183-EN-N, ISBN: 978-92-846-0286-5 (paper), ISBN: 978-92-846-0288-9 (pdf), DOI: 10.2861/94270 (paper), DOI: 10.2861/094050 (pdf). Downloadable from http://meszerics.eu/pdf/energypovertyhandbook-online.pdf.

United Nations, *Universal Declaration of Human Rights*. http://www.un.org/en/universal-declaration-human-rights/.

United Nations, *Resolution 64/292: The Human Right to Water and Sanitation*. A/RES/64/292, 3rd August 2010. http://www.un.org/es/comun/docs/?symbol=A/RES/64/292&lang=E.

United Nations, *2014 UN-Water Annual International Zaragoza Conference*. Preparing for World Water Day 2014: Partnerships for improving water and energy access, efficiency and sustainability. 13–16 January 2014. http://www.un.org/waterforlifedecade/water_and_energy_2014/side_breakfast_legal_tenure_aspects_water_energy.shtml.

World Health Organization (WHO), *What Is the Minimum Quantity of Water Needed?* http://www.who.int/water_sanitation_health/emergencies/qa/emergencies_qaen/.

World Health Organization (WHO), *Technical Notes On Drinking-Water, Sanitation and Hygiene in Emergencies*, Updated: July 2013, http://www.who.int/water_sanitation_health/publications/2011/WHO_TN_09_How_much_water_is_needed.pdf?ua=1.

2

General Theory of Wind-Driven Machines

2.1 Betz's Theorem

The German engineer and physicist Albert Betz was the first to scientifically study the operation principle of wind turbines, determining through a series of reasoning what is the maximum power available from an open fluid stream, i.e., air or water flowing in a space not confined by walls. Betz's Theorem is as fundamental for understanding wind-driven machines as Carnot's theorem is for thermal machines.

Betz's theory is based on four postulates:

a. The airflow across an imaginary section of a turbine, S_1, placed upwind of its rotation plane, has a speed V_1.

b. The airflow across an imaginary section, S_2, placed downwind of its rotation plane, has a speed V_2 (see Figure 2.1).

c. The fluid is incompressible and its speed is constant throughout the section S.

d. The turbine is an "actuator disk" of infinitesimal thickness and has no friction losses, nor does it modify the flow's direction. In reality, the flow downwind of the turbine is vortical, a fact that complicates much of the analysis. Hence, Betz's simplifying assumption is that the rotation of the turbine does not transmit any tangential motion to the fluid. In short, the flow is bi-dimensional and its only components are axial and radial.

Because of the *Energy Conservation Principle*, if a turbine extracts a certain amount of energy from an airflow, the latter must transfer the same amount of kinetic energy. Hence, the speed V_2 must be slower than the speed V_1. If we assume that air is incompressible, then the following equation of continuity must be satisfied:

$$S_1 \cdot V_1 = S_2 \cdot V_2 = S \cdot V$$

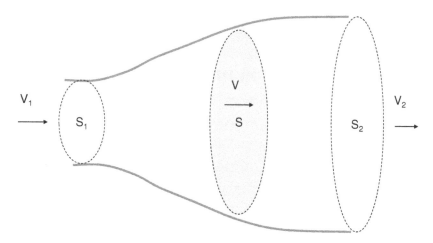

FIGURE 2.1
The motion of a fluid across an ideal turbine.

Please note that the product of an area by a speed is a volume per unit of time (m² · m/s = m³/s). Hence, in the continuity equation the volume of air flowing across a given area is constant, because of the assumption that air is incompressible.

In order to satisfy the said equation, the section S_1, corresponding to the flow entering the turbine, must be smaller than section S_2, corresponding to the flow leaving the turbine.

The section S indicates the area swept by the rotor.

If we calculate the force (F) applied by the turbine on the airflow, according to *Euler's Equation* its absolute value results from the following expression:

$$F = Q \cdot (V_1 - V_2) = S \cdot V \cdot \rho \cdot (V_1 - V_2)$$

where:
  F = Absolute value of the force (in forward direction) (N)
  ρ = Density of air (1.2–1.25 kg/m³, depending on several factors)
  Q = Airflow (kg/s)
  V_1 = Speed of the air upwind of the turbine (m/s)
  V_2 = Speed of the air downwind of the turbine (m/s)
  V = Speed of the air as it flows across the turbine (m/s)

The power P absorbed by the turbine, is the product of the force F by the speed V. Please note that, since motion is relative, it is equivalent to consider a static turbine and the air in motion, or the air as being static and the turbine moving with the said speed. The power is then:

$$P = F \cdot V = \rho \cdot S \cdot V^2 \cdot (V_1 - V_2) \qquad (2.1)$$

If we now consider the power absorbed by the turbine as the variation of the kinetic energy contained in the mass of air (ΔE) passing through the section S of the turbine in each second, we will get the following expression:

$$\Delta E = \frac{1}{2} \cdot \rho \cdot S \cdot V \cdot (V_1^2 - V_2^2) = P \tag{2.2}$$

By equaling Equations 2.1 and 2.2, we get:

$$P = \rho \cdot S \cdot V^2 \cdot (V_1 - V_2) = \frac{1}{2} \cdot \rho \cdot S \cdot V \cdot (V_1^2 - V_2^2)$$

$$V \cdot (V_1 - V_2) = \frac{1}{2} \cdot (V_1^2 - V_2^2) = \frac{1}{2} \cdot (V_1 - V_2) \cdot (V_1 + V_2)$$

Simplifying the last one, we derive the following important conclusion:

$$V = \frac{V_1 + V_2}{2} \tag{2.3}$$

If we now replace Equation 2.3 with Equation 2.1, we get the following expression:

$$P = \frac{1}{4} \cdot \rho \cdot S \cdot (V_1 + V_2) \cdot (V_1^2 - V_2^2) \tag{2.4}$$

If we now assume that V_1 is constant, it is then possible to calculate the value of V_2 that maximizes the power P, by deriving Equation 2.4, equaling the derivative to zero, and calculating the roots, as follows:

$$\frac{dP}{dV_2} = \frac{1}{4} \cdot \rho \cdot S \cdot (V_1^2 - 2 \cdot V_1 \cdot V_2 - 3 \cdot V_2^2) = 0$$

The former quadratic equation has two roots:
$V_2 = -V_1$ (physically impossible)
$V_2 = V_1/3$ (corresponding to the maximum value of the power)

If we introduce the said value of V_2 in the expression of P, we obtain the value of the maximum power extractable from the open flow:

$$P_{max} = \frac{8}{27} \cdot \rho \cdot S \cdot V_1^3 = \frac{16}{27} \cdot \left(\frac{1}{2} \rho \cdot S \cdot V_1^3 \right)$$

Such expression, known as *Betz's formula,* states practically that the maximum power that is possible to extract from an open stream of any fluid is equal to 59.25% of the total kinetic power contained in the undisturbed flow upwind of the turbine.

2.2 The Extension of Betz's Theorem to Vertical Axis Wind Turbines

In a vertical axis turbine, half of the blades are always downwind with respect to the other half. One could suppose then that a vertical axis turbine can be analyzed as if it was a couple of actuator disks, placed in cascade along the same stream of fluid. Let us assume that each actuator disk turns independently from the other, so that their exposed areas are both equal to S and that no interference between them exists. Mathematically speaking, this means that the speeds V_1, V_m, and V_2 are all independent from each other. Furthermore, it is assumed that the fluid is incompressible, and hence its density is constant along the whole trajectory through both actuator disks. Figure 2.2 illustrates these suppositions.

Applying to the actuator disk 1 the reasoning already explained in the former Section 2.1, we obtain:

$$V' = \frac{V_1 + V_m}{2}$$

where:
V_1 = Speed of the fluid upstream of the actuator disk 1 (m/s)
V_m = Speed of the fluid downstream of the actuator disk 1 (m/s)
V' = Speed of the fluid passing through the actuator disk 1 (m/s)

Now let us express the speed V' as a fraction of the speed V_1, given from the following equation:

$$V' = V_1(1-\alpha)$$

in which α is called "speed drop coefficient" or "interference coefficient." Equaling both expressions of V', we get:

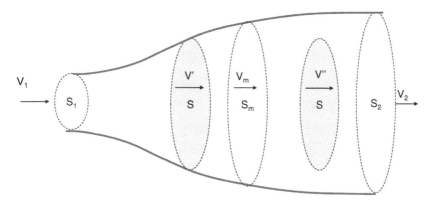

FIGURE 2.2
Hypothesis for the extension of Betz's theorem to two actuator disks disposed in series.

$$V_m = V_1(1 - 2\alpha)$$

If we now apply Equation 2.2 to the actuator disk 1, the result is:

$$\Delta E = \frac{1}{2} \cdot \rho \cdot S_1 \cdot V' \cdot (V_1^2 - V_m^2) = P$$

Replacing V_m and V' with their equivalents as a function of V_1, we will obtain the power extracted from the fluid by the actuator disk 1:

$$P_1 = \frac{1}{2} \cdot \rho \cdot S_1 \cdot V' \cdot \left[V_1^2 - V_1^2(1 - 2a)^2 \right]$$

$$P_1 = \frac{1}{2} \cdot \rho \cdot S_1 \cdot V' \cdot V_1^2 \cdot 4a \cdot (1 - a)$$

$$P_1 = \frac{1}{2} \cdot \rho \cdot S_1 \cdot V_1^3 \cdot 4a \cdot (1 - a)^2$$

Applying now the same reasoning to the actuator disk 2, we can write:

$$V'' = V_m(1 - b)$$

where b is the interference coefficient of the second actuator disk and then:

$$V_2 = V_m(1 - b)$$

In the same way as already deduced in Section 2.1, the power extracted from the flow by the second actuator disk results from the following equation:

$$P_2 = \frac{1}{2} \cdot \rho \cdot S_2 \cdot V'' \cdot \left[V_m^2 - V_m^2(1 - 2b)^2 \right]$$

Now replacing the expression of V'' in the equation of P_2, after some simple algebraic steps, we will get the following expression:

$$P_2 = \frac{1}{2} \cdot \rho \cdot S_2 \cdot V_m^3 \cdot 4 \cdot (b - 2b^2 + b^3)$$

Since the speed V_m depends on the actuator disk upwind, and we have supposed that the actuator disk downwind does not induce any effect on it, the power extracted by the second actuator disk will be hence a function of the single variable b, because all other factors result to be constant. Deriving the expression of P_2 as a function of b and equaling the derivative to zero, we get the following result:

$$1 - 4b + 3b^2 = 0$$

The only valid solution for the equation above is then:

$$b = \frac{1}{3}$$

Consequently, the value of V_m is:

$$V_m = 3V_2$$

With the same reasoning, the power P_1 depends only on the variable a, hence the said power will be maximum when its derivative as a function of a is null. As in the former case, such condition is satisfied when:

$$a = \frac{1}{3}$$

and hence:

$$V_m = \frac{V_1}{3}$$

The maximum total power that can be extracted by two actuator disks in a fluid stream is then:

$$P_{max} = P_1 + P_2 = \frac{1}{2} \cdot \rho \cdot S_1 \cdot V_1^3 \cdot 4a \cdot (1-a)^2 + \frac{1}{2} \cdot \rho \cdot S_2 \cdot V_m^3 \cdot 4 \cdot (b - 2b^2 + b^3)$$

$$P_{max} = \frac{1}{2} \cdot \rho \cdot S \cdot V_1^3 \cdot \left[4a \cdot (1-a)^2 + \frac{4}{9} \cdot (b - 2b^2 + b^3) \right]$$

$$P_{max} = \frac{1}{2} \cdot \rho \cdot S \cdot V_1^3 \cdot \frac{16}{27} \cdot \left(1 + \frac{1}{9} \right)$$

The expression above states that the power from two actuator disks placed in series along a fluid stream is bigger than the power that a single actuator disk can extract, and equal to 65.8% of the power contained in the fluid upstream of the first actuator disk.

2.2.1 Discussion of the Extension of Betz's Theorem to Vertical Axis Turbines

We will see at the end of this chapter that, in practice, vertical axis turbines always feature lower aerodynamic efficiency than horizontal axis turbines. The reason for such apparent contradiction with the extension of Betz's Theorem is that the blades' path of a vertical axis turbine is circular; consequently, the trajectory of the blade is orthogonal to the direction of the wind in only two points. In the supposition that the vertical axis turbine is equivalent to two actuator disks in series, we are implicitly assuming that the trajectory of the blades is in any instant orthogonal to the direction of the wind hence, we assume that the path of the blades is square instead of circular. We can, therefore, conclude that the calculation model employed for the extension of Betz's Theorem does not represent accurately the operation of a vertical axis turbine, but instead it represents well the situation in which two horizontal axis turbines are placed in series along an airflow. In such case, each turbine behaves like an independent actuator disk and the hypothesis formulated at the beginning of Section 2.2 is acceptably close to the reality.

It is important to remark that, from a practical point of view, the eventual duplication of the rotors of a horizontal axis wind turbine would almost double the production cost, but the increase in its efficiency would be modest (11%) compared to a horizontal axis wind turbine having a single rotor.

2.3 Notions on the Theory of Wing Sections

The principal component of a wind-driven machine, regardless of being a modern turbine or an ancient windmill, is the blade. In practice, wind turbines' blades are nothing but rotating airplane wings. In order to understand their operation and for their correct dimensioning, it is necessary to introduce some notions on the theory of wing sections, a.k.a. airfoils. Indeed, Betz's Theorem does not provide any information about how to build a blade.

Consider then an aerodynamic profile surrounded by an airflow of speed V, and let us define some magnitudes (please refer to Figure 2.3).

Definitions

Leading edge: Is the set of points A.

Trailing edge: Is the set of points B.

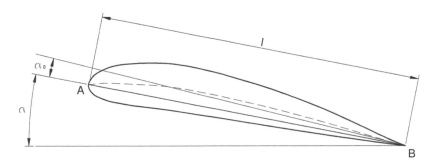

FIGURE 2.3
Typical section of an aerodynamic profile.

Line AB: Chord of the profile.

Dotted line: Called camber line, is the geometric place of the middle points between the extrados and the intrados

l: Length of the chord.

α: Angle of attack (a.k.a. pitch angle) of the profile. It is the angle between the chord and the direction of the wind.

α_0: It is the angle of attack at which the profile's lift becomes null. In symmetric profiles, such angle is equal to zero, while in asymmetric profiles it is a negative value.

In general, airfoils are classified as concave–convex, biconvex, and flat-convex. The lower side with reference to the leading edge is called intrados, and the upper side, extrados.

Furthermore, airfoils are classified according to their relative thickness or fineness coefficient, defined as:

$$e = \frac{h}{l}$$

where:
h = Maximum thickness of the airfoil
l = Chord of the airfoil

When e is smaller than 6%, between 6% and 12%, and bigger than 12%, the airfoil is, respectively, said to be thin, semi-thick, and thick.

In general, the aerodynamic efficiency, intended as quotient between lift and drag of the airfoil, is inversely proportional to the fineness coefficient.

The geometric data of airfoils are conventionally provided as percentage tables, where the origin of the Cartesian axes coincides with the leading edge and the chord is by definition equal to 100% (or equal to 1), coincident with the abscises axis. The ordinates are defined as a percentage of the chord's length too. In this way, it is very easy to scale an airfoil, just by defining the length of its chord, l. Generally, the geometric data are tabulated as a vector, starting with the coordinate 1 and proceeding until 0, then from 0 to 1 (Selig format). In other cases, they are defined as two separate vectors, one defining the coordinates of the extrados from the leading edge until the trailing edge, and the other vector providing the coordinates of the intrados (Lednicer format). In some cases, the literature on airfoils provides a third vector of coordinates, corresponding to the camber line.

Table 2.1 shows an example of the geometric definition of an airfoil, taken from a database for aircraft models, which is based on the first convention (Selig format). Figure 2.4 shows the graphic result, obtained with a free online plotter. Please observe that the plotter provides the camber line too, which is useful for finding the center of pressure, called *CP* and defined in Section 2.3.

TABLE 2.1

Geometry of the Airfoil NREL S 820

Name	NREL's S820 Airfoil
Chord (mm)	100
Radius (mm)	0
Thickness (%)	16
Origin (%)	0
Pitch (°)	0

Airfoil Surface

X (mm)	Y (mm)
100	0
99.6214	0.0505
98.5214	0.254
96.7892	0.6537
94.5228	1.239
91.7963	1.964
88.6485	2.7777
85.0998	3.6689
81.2173	4.6401
77.0798	5.6654
72.7639	6.703
68.3412	7.6946
63.8577	8.5468
59.2824	9.2241
54.6545	9.7201
49.9848	10.0066
45.2681	10.1025
40.5512	10.0437
35.8861	9.8462
31.3252	9.5223
26.9179	9.0832
22.7135	8.5402
18.7577	7.9047
15.0957	7.1882
11.7679	6.4007
8.8084	5.5543
6.2503	4.6603
4.1136	3.7327
2.4243	2.7843
1.1711	1.8325
0.365	0.9309
0.0127	0.1407
0.0109	0.1293

(Continued)

TABLE 2.1 (*Continued*)

Geometry of the Airfoil NREL S 820

Name	NREL's S820 Airfoil
0.0008	−0.0335
0.0091	−0.1066
0.0286	−0.1719
0.0646	−0.2321
0.1162	−0.2929
0.246	−0.409
0.2568	−0.4173
1.2145	−0.9433
2.7631	−1.5109
4.8455	−2.0953
7.4139	−2.6826
10.4315	−3.2597
13.8524	−3.8136
17.6342	−4.3316
21.7272	−4.8005
26.0834	−5.2072
30.6489	−5.5382
35.3714	−5.7789
40.1939	−5.9126
45.0609	−5.9167
49.9228	−5.7439
54.7944	−5.3587
59.6957	−4.8001
64.5793	−4.1049
69.4524	−3.3172
74.2551	−2.5276
78.9038	−1.7945
83.3084	−1.1608
87.3751	−0.6548
91.0106	−0.29
94.126	−0.0641
96.641	0.0405
98.4887	0.0559
99.6194	0.0246
100	0
Camber Line	
X (mm)	Y (mm)
0.0109	0.1293
0.0008	−0.01675
0.0091	−0.0533

(*Continued*)

TABLE 2.1 (*Continued*)

Geometry of the Airfoil NREL S 820

Name	NREL's S820 Airfoil
0.0286	0.002232
0.0646	0.012505
0.1162	0.039974
0.246	0.127493
0.2568	0.135455
1.2145	0.461081
2.7631	0.731804
4.8455	0.977569
7.4139	1.192176
10.4315	1.379399
13.8524	1.540191
17.6342	1.676639
21.7272	1.790625
26.0834	1.884112
30.6489	1.95836
35.3714	2.015374
40.1939	2.057987
45.0609	2.091609
49.9228	2.13198
54.7944	2.173203
59.6957	2.181409
64.5793	2.152371
69.4524	2.064131
74.2551	1.908448
78.9038	1.70945
83.3084	1.478108
87.3751	1.221347
91.0106	0.938551
94.126	0.640206
96.641	0.366236
98.4887	0.158723
99.6194	0.037735
100	0
Chord Line	
X (mm)	Y (mm)
0	0
100	0

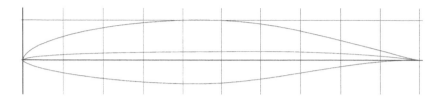

FIGURE 2.4
Graphical representation of the airfoil NREL S820, in scale, obtained with the free online
plotter of www.airfoiltools.com.

2.4 Action of the Air on a Wing in Motion

Experience shows that, when placing a profiled wing in an air stream,
inclined with a certain angle in relation to the direction of the latter, the
intrados will suffer an overpressure while the extrados will be in depression.
The wind's action on the wing manifests itself as a resultant force F, generally oblique to the direction of the relative air speed, which we will note
with V. Such force, called total aerodynamic action or just action of the air,
can be calculated by means of the following expression:

$$F = \frac{1}{2} \cdot \rho \cdot C_t \cdot V^2 \cdot S$$

where:
  C_t = Dimensionless coefficient, called total aerodynamic coefficient
  ρ = Density of the air
  S = Planform area of the wing

2.4.1 Lift, Drag, and Moment Coefficients

In general, it is not important to know the force F, but its projections on a
system of reference axes, x and z, coincident with the relative speed V and
its normal.
  The following parameters are then defined:

  C_z = Lift coefficient, sometimes called C_l in the literature
  C_x = Drag coefficient, sometimes called C_d

The lift and drag forces can be calculated with the following formulas:

$$F_x = \frac{1}{2} \cdot \rho \cdot C_x \cdot V^2 \cdot S$$

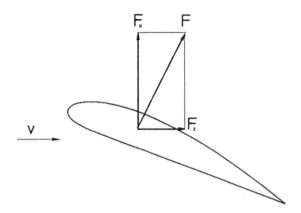

FIGURE 2.5
The forces acting on an airfoil.

$$F_z = \frac{1}{2} \cdot \rho \cdot C_z \cdot V^2 \cdot S$$

Figure 2.5 shows the distribution of the forces on the wing.
 By Pythagoras' Theorem:

$$F^2 = F_x^2 + F_z^2$$

Replacing the expression of each force and simplifying, we get:

$$C_t^2 = C_x^2 + C_z^2$$

Another parameter useful for the design is the coefficient of moment C_m defined by the following formula:

$$M = \frac{1}{2} \cdot \rho \cdot C_m \cdot V^2 \cdot S \cdot l$$

where M is the moment of the aerodynamic force to the trailing edge, and l is the chord of the airfoil.
 The position of the center of pressure, CP, relative to the trailing edge, x_1, results from the following expression:

$$CP = \frac{x_1}{l} = \frac{C_m}{C_z}$$

2.4.2 Graphical Representation of the Aerodynamic Coefficients C_x and C_z

Since the relationships between α, C_z, C_x, and C_m vary in a very complex way among different airfoils, and such relationships are difficult to model with analytical formulas, it is common practice in the literature to provide the airfoil's aerodynamic features in graphical form. The resulting curves are generically called *polars*. There are three modes of plotting polars:

- As Cartesian curves (which, by definition, are not polar curves!);
- As *Eiffel's polar* (so-called in honor of the famous French engineer, who was the first to measure the aerodynamic features of objects in a wind tunnel, which is still in operation);
- As *Lilienthal's polar* (seldom employed in modern aerodynamics, but useful for the design of vertical axis wind turbines of the Darrieus type), and finally;
- As mixed Cartesian—Polar curves.

2.4.2.1 Cartesian Representation of C_x, C_m, and C_z as a Function of the Pitch Angle

Plotting the values of C_z and C_x as a function of α (Figure 2.6) is the most straightforward criterion for graphical representation.

2.4.2.2 Eiffel's Polar

Such curve employs the coefficient of aerodynamic drag, C_x as abscises, and the coefficient of lift, C_z as ordinates. The said curve is generally graduated as a function of the pitch angle, or combined with the Cartesian curves. Figure 2.7 shows the polar of the airfoil E371.

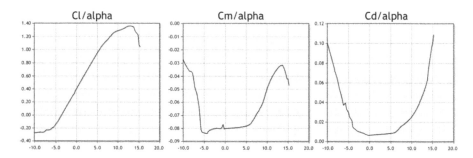

FIGURE 2.6
An example of Cartesian representation of an airfoil's aerodynamic coefficients.

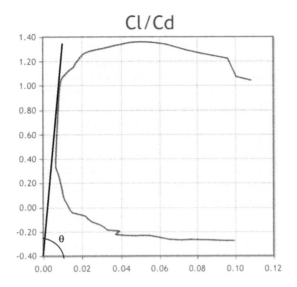

FIGURE 2.7
Polar of the airfoil E371, plotted with the free online aerodynamic simulator and plotter of airfoils www.airfoiltools.com.

The ratio between lift and drag is often referred to in the literature as *glide coefficient,* or *aerodynamic efficiency coefficient,* or *lift do drag ratio (L/D ratio),* or *fineness ratio, f.* The coefficient *f* is usually plotted as a function of the angle θ. In the Eiffel's polar, *f* is maximum when the straight line with center in the origin of the axes is tangent to the polar.

$$f = \frac{C_z}{C_x} = \tan\theta$$

2.4.2.3 Lilienthal's Polar

This is actually a Cartesian graph, in which the coefficients represented are not C_z and C_x, but their projections perpendicular and parallel to the chord of the airfoil, respectively, C_r and C_t. The relationship of both systems of coefficients with the pitch angle is given by a pair of equations:

$$C_t = C_z \cdot \mathrm{sen}\,\alpha - C_x \cdot \cos\alpha$$
$$C_r = C_z \cdot \cos\alpha + C_x \cdot \mathrm{sen}\,\alpha$$

2.4.2.4 Mixed Representations

These are more frequent in books and other sources presenting data measured in the wind tunnel.

2.4.3 Definitions and Terminology

Before starting the study of the different types of wind turbines, it is useful to define a series of dimensionless coefficients that will turn useful in order to compare turbines that otherwise would not be comparable. They are also useful to calculate mechanical loads, the speed of rotation, and other parameters of operation that usually are not known in advance.

2.4.3.1 Solidity Coefficient, σ

The solidity coefficient is defined as the ratio between the area of the blades and the hub (solid colored area in Figure 2.8) and the area swept by the rotor (filled area in Figure 2.8).

$$\sigma = \frac{a}{A}$$

In general, wind-driven machines with high solidity feature high torque and low speed of rotation; they are then useful for driving positive displacement water pumps. On the contrary, low solidity turbines present opposite features and are more suitable to drive centrifugal pumps or electrical generators.

2.4.3.2 Specific Speed, λ

It is the ratio between the tangential speed of the blade's tip and the speed of the wind.

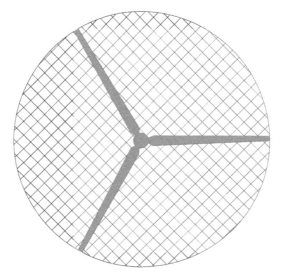

FIGURE 2.8
Graphic representation of the solidity coefficient.

$$\lambda = \frac{\omega \cdot R}{V} = \frac{\pi \cdot D \cdot N}{60 \cdot V}$$

where:

ω = Angular speed (rad/s)
N = Angular speed (R.P.M or turns/min)
V = Speed of the wind (m/s)
$R = D/2$ = Radius (m)

2.4.3.3 Coefficient of Motor Torque, C_M

C_M (not to be mistaken with the coefficient of moment, C_m, already defined in Section 2.3.1) is the ratio:

$$C_M = \frac{2 \cdot M}{\rho \cdot S \cdot R \cdot V^2}$$

where:

M = Motor torque (Nm)
S = Area swept by the blades (m²)
V = Speed of the wind (m/s)
$R = D/2$ = Radius of the rotor (m)
ρ = Density of air (kg/m³)

2.4.3.4 Coefficient of Axial Force, C_F

C_F is defined as the ratio:

$$C_F = \frac{2 \cdot F}{\rho \cdot S \cdot R \cdot V^2}$$

where:

F = Axial force (N)
S = Area swept by the blades (m²)
V = Speed of the wind (m/s)
$R = D/2$ = Radius of the rotor (m)
ρ = Density of air (kg/m³)

2.4.3.5 Coefficient of Power, C_P

The coefficient of power C_P (not to be mistaken with the pressure center CP already defined in Section 2.3.1) is given by the ratio:

$$C_P = \frac{2 \cdot P}{\rho \cdot S \cdot V^3}$$

where:
P = Power of the turbine (W)
S = Area swept by the blades (m²)
V = Speed of the wind (m/s)
ρ = Density of air (kg/m³)

2.4.3.6 Relationships between Dimensionless Coefficients

From the definitions presented above it is possible to derive the following expressions:

$$F = \frac{\rho}{2} \cdot C_F \cdot S \cdot V^2$$

$$P = \frac{\rho}{2} \cdot C_P \cdot S \cdot V^3$$

$$M = \frac{\rho}{2} \cdot C_M \cdot R \cdot S \cdot V^2$$

$$N = \frac{\lambda \cdot V}{2 \cdot \pi \cdot R} = \frac{\lambda \cdot V}{\pi \cdot D}$$

The said relationships are fundamental to consent comparisons between the different types of turbines, by plotting the curves of the coefficients C_M, C_P, and C_F as a function of the specific speed λ. In this way, it is possible to represent the mechanical properties of F, M, P, and N as a function of the wind speed, V, with N as parameter.

We can easily demonstrate the relationships between the dimensionless coefficients C_F, C_P, and λ.

If we start by the definition of mechanical power:

$$P = M \cdot \omega$$

and replace by the equivalent expressions:

$$\frac{\rho}{2} \cdot C_P \cdot S \cdot V^3 = \frac{\rho}{2} \cdot C_M \cdot R \cdot S \cdot V^2 \cdot \lambda \cdot \frac{V}{R}$$

we get then:

$$C_P = C_M \cdot \lambda$$

In the same way:

$$P = F \cdot V \quad \text{(by definition)}$$

hence,

$$\frac{\rho}{2} \cdot C_P \cdot S \cdot V^3 = \frac{\rho}{2} \cdot C_F \cdot S \cdot V^2 \cdot V$$

$$C_P = C_F$$

2.4.3.7 Reynolds' Number

This dimensionless number is a measure of the turbulence in a fluid stream. It is a fundamental parameter for the design of small wind turbines, especially in sites featuring weak winds, because the aerodynamic performances of airfoils sink, and in some airfoils even become erratic, below a given Reynolds' number.

The Reynolds' number, Re, is defined as:

$$\mathrm{Re} = \frac{V \cdot d}{\nu}$$

where

ν = Kinematic viscosity of air $\approx 14.607 \cdot 10^{-6}$ (m^2/s) (N.B. it varies with P and T of air)

d = Characteristic distance (m)

The characteristic distance d, is the principal dimension of the object related to the Re, measured in the sense of the fluid stream. For instance, the Re of an airfoil is calculated taking as distance d the chord of the airfoil; the Re of a tubular tower is calculated assuming as distance d of the diameter of the tube.

If Re < 10^7 then the flow is said *laminar*, and the fluid flows adherent to the surface of the object under analysis. If Re > 10^9 then it is said turbulent and it is possible to observe the fluid detaching from the surface of the object under analysis. Intermediate values of Re generate intermediate aerodynamic behaviors, because the fluid flows adhering in part, and then detaching from the surface of the object.

2.5 Classification of Wind Turbines

Two big families of wind turbines exist: those featuring a horizontal axis and those having a vertical axis. Each family, in turn, is composed of different types. Furthermore, there is a long list of wind-driven machines, not always definable as "turbines," and often of absurd conception.

Each family and type has its specific characteristics, advantages, and disadvantages, as we will see later on.

2.5.1 Vertical Axis Wind Turbines

This type of turbine is classified into three families: reaction-driven turbines, aerodynamic action turbines, and mixed or hybrid turbines.

2.5.1.1 Reaction-Driven Turbines

The classical example is the pinwheel, with which all of us have played as children. This type of wind turbines is technically called *pannemones*, (from ancient Greek *pan*, παν, every, and πνευμον, *pneumon*, air) because it will turn regardless of the direction from which the wind is coming. The most diffused type is the Savonius rotor that will be studied in detail in Chapter 6.

2.5.1.2 Aerodynamic Action Turbines

The classical example is the Darrieus rotor, developed in France between both World Wars. In the last years, there is a growing interest in the public for the Gorlov rotor. The latter is just a Darrieus rotor in which the blades feature a certain torsion that makes it aesthetically more appealing, and provides some positive starting torque. The simplified design of Darrieus rotors with straight blades will be treated in detail in Chapter 5.

2.5.1.3 Hybrid Turbines

These turbines start their operation because of the differential reaction forces and, as their speed of rotation grows, they trigger some sort of mechanism of aerodynamic lift generation. The most diffused type is the *Lafond* turbine, a pneumatic version of hydraulic turbines, and the *Gyromill®* rotor, developed and patented by the MIT *(Massachussets Institute of Technology)* during the petrol crisis of the 1970s, purchased by a division of the airplane manufacturer *McDonnel-Douglas®*, which in turn failed in the 1990s. Another type of mixed action wind turbine is the *Cycloturbine*, patented around the middle of the nineteenth century, and periodically "resurging" in Internet blogs and e-zines.

In the present volume, we will not deal with the design of *Lafond turbines*, *Gyromill®*, or *Cycloturbine*, because they feature mediocre performance and their construction is rather complicated.

2.5.2 Horizontal Axis Wind Turbines

Horizontal axis wind turbines are classified into two families: slow and fast turbines.

2.5.2.1 Fast Turbines

The number of blades of this type of wind-driven machines varies from two to four, most often three. This family of turbines features low values of solidity coefficient σ (in general about 5%) and high specific speed λ (between 5 and 10). Nevertheless, single-blade turbines have been proposed that feature even higher specific speeds, but pose serious construction challenges and instability issues (from the 1970s though the interest in them has fallen). The typical use of fast wind turbines is the generation of electricity. The advantage of fast wind turbines is that, for a given rated power, they are lighter than slow wind turbines because of their lower solidity coefficient. Their main drawback is a low starting torque; in practice, they need wind speeds greater than 5 m/s in order to run stably, because at lower wind speeds, the Re is low, and the aerodynamic performance of the airfoils is mediocre. The blades can be built with fixed or variable pitch, with or without torsion, with upwind or downwind rotor (respectively, oriented by means of a rudder, or by rotor eccentricity/conicality, or in other cases by means of servomotors and wind direction sensors). The blades can be manufactured in wood, aluminum, or fiber-reinforced resins. It is possible to also build sail rotors, although such type is not commercially diffused, being suitable for "do-it-yourself" models. Figure 2.9 shows some constructive sketches and Figure 2.10 illustrates the different types of blades.

Fast turbines have better aerodynamic performances than slow turbines do. Figure 2.11 illustrates the variation of the dimensionless coefficients C_M and C_P as a function of the specific speed λ, measured experimentally on a turbine with 3 kW rated power running with fixed pitch and variable speed (Bunlung et al. 2010). Please note that the maximum CP, i.e., the maximum efficiency, corresponds to values of λ comprising between 6 and 7. This is a common feature to small fast turbines.

If we assume:

$$\lambda = \frac{\pi \cdot D \cdot N}{60 \cdot V} = 6$$

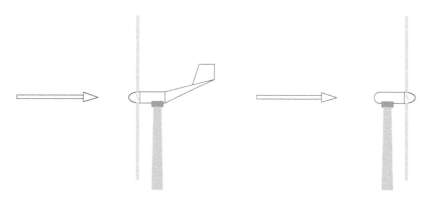

FIGURE 2.9
Upwind and downwind rotors.

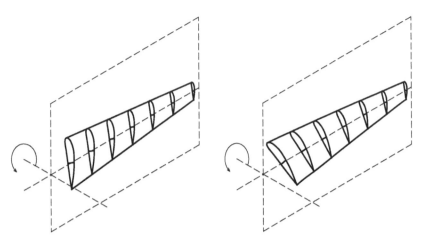

FIGURE 2.10
Flat blade (left) and twisted blade (right). The dotted parallelogram indicates the rotation plane and the round arrow the rotation sense.

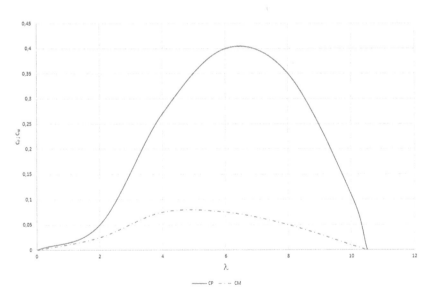

FIGURE 2.11
C_M and C_P as a function of λ. Data from Bunlung et al., graphics by the Author.

by replacing all known constants, we obtain a practical formula for predefining the speed of rotation:

$$N = \frac{115 \cdot V}{D}$$

where:
N (R.P.M.), V (m/s), D (m).

The expression above, together with the following formula, derived from Betz's Theorem, allows the designer to get a preliminary approximation of the features of the turbine (diameter and speed of rotation) by defining just the desired power and the most frequent speed of the wind in the site:

$$P = 0.20 \cdot D^2 \cdot V_1^3$$

Another advantage of fast wind turbines on slow ones is the low wind load on the hub when the rotor is stopped, because such load is smaller than the one arising from the normal operation conditions. The explanation is simple. Since

$$C_P = C_F$$

when the rotor is stopped it does not produce energy, and hence $C_P = 0$ and consequently $C_F = 0$. In practice, the force acting on a blocked rotor is not null. In first approach, we can estimate that the exposed area of each blade is a flat rectangle. Since the solidity, σ, of quick turbines is about 5% of the swept area, the wind pressure resulting on the rotor's hub is minimum.

2.5.2.2 Slow Wind Turbines

These turbines appeared first in the USA around 1870, diffusing from there to the rest of the world. For instance, the *National Directorate for Energy Conservation and New Sources* of Argentina estimated at the end of the 1980s that around one million of windmills of such type were installed and operative in the Province of Buenos Aires for pumping water. If we consider that such turbines yield in average 300 W with the winds of the zone, the average power employed for pumping water for agriculture is 300 MW, the equivalent of a small thermoelectric plant continuously running all year round. This datum is the base for a reflection: wind power does not need to be limited to electricity generation, as it has been the general trend in the last 20 years. We are going to see in Chapter 10 that the generation of electricity is not always the most sustainable solution, since water storage is cheaper and easier to implement than energy storage.

Slow wind turbines are very suitable for pumping water. Windmills are much diffused in developing countries, and all commercial models are just copies with small variations of the American type developed in 1870.

Slow wind turbines feature high solidity σ, and high starting torque hence, they are able to start with wind speeds as weak as 2 or 3 m/s, even under load. The number of blades ranges from 12 to 24. In general, these latter are just curved metal plates and because of this, their aerodynamic performance is always low. Employing profiled blades would mean an increase of the manufacturing cost, so the strategy of practically all manufacturers is to increase the diameter of the rotor rather than to increase its efficiency.

FIGURE 2.12
Appearance of the multi-blade turbine for pumping water. © Steve Landgrove, www.dreamstime.com, rights purchased by the Author.

Figure 2.12 illustrates the general appearance of the so-called "American multi-blade" windmill. The curve of variation of C_P and C_M as a function of λ shown in Figure 2.13 is the result of tests performed in the *Eiffel Laboratory* of Paris, published by *Le Gourière*, the graphic elaboration is that of the Author's original book of 1992.

The efficiency of these turbines is maximum when $\lambda \approx 1$.

In such conditions, the speed of rotation is:

$$N = \frac{60 \cdot V}{\pi \cdot D} = 19 \frac{V}{D}$$

Typically, their coefficient of power, C_P, is around 0.3. Hence, the resulting aerodynamic efficiency is 50% of the maximum value predicted by Betz's Theorem. Based on this consideration, we can derive the following formula:

$$P = 0.15 \cdot D^2 \cdot V_1^3 \rightarrow D = \sqrt{\frac{P}{0.15 \cdot V_1^3}}$$

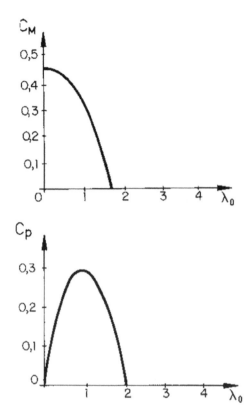

FIGURE 2.13
Characteristic operational curves of the American multi-blade, according to *Le Gourière*, graphics by the Author.

Both formulas turn useful for pre-dimensioning the rotor.

Turbines of the American multi-blade type usually lack a system of automatic switch-off in case of strong wind, which makes them very vulnerable to heavy damage if the owner does not manually fold them out before storms, placing the rotor in the "flag position." Figure 2.14 shows the kind of damage that strong winds can cause if the windmill is not protected. The rotor of these windmills is in general so heavy, that their diameters usually do not exceed 8 m, though in the USA and China some models reached 15 m of diameter.

2.5.3 "Undefinable" Wind-Driven Machines

Nothing like capturing energy from the wind has ever stimulated the fantasy of "inventors" in all times. The capillarity reached by Internet and the availability of 3D solid modeling software has boosted any kind of hoaxes about wind-driven machines capable of "cheating" the Betz's

FIGURE 2.14
Damage caused to a multi-blade turbine during a storm in Australia. © Carol Hancock, www.dreamstime.com, rights purchased by the Author.

Theorem. The analysis of "undefinable" wind-driven machines, i.e., those that are not "turbines," or are combinations of conventional turbines with any kind of accessories, would require a book in itself. A few examples will be treated in detail in Chapter 11. For the moment we can just propose to the reader a simple rule: since the Betz's Theorem is a limit law of Physics—by the way, based on very optimistic postulates—in practice, it is impossible to extract more energy from a free-flowing stream than its theoretical maximum.

2.5.4 Comparison between Different Types of Wind Turbines

Figure 2.15 shows the comparison between the aerodynamic efficiencies of different types of wind turbines. The correct way to compare turbines having different sizes, and even different operation principles, is by employing dimensionless coefficients, i.e., plotting C_P as a function of λ. Figure 2.15 provides a panoramic view of the aerodynamic efficiency featured by each family of turbines. Figure 2.16 shows the theoretical influence of the number of blades on the aerodynamic efficiency of the rotor.

Both figures allow us to extract the following conclusions that anticipate the detailed analysis in Chapter 3:

a. Horizontal axis wind turbines are more efficient than vertical axis wind turbines, because their operation is better applicable to Betz's postulates, while vertical axis turbines, because of their own

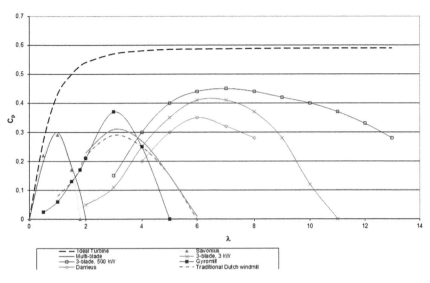

FIGURE 2.15
Comparison between different types of wind turbines. Experimental data from several sources, graphics by the Author.

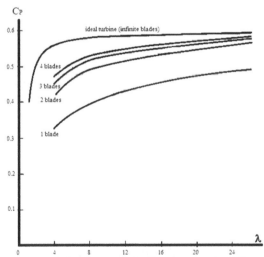

FIGURE 2.16 Comparison of the theoretical efficiencies with different numbers of ideal (dragless) blades, according to *Le Gourière*, graphics by the Author.

nature and geometry, induce intense vortexes that dissipate energy and occupy a "volume" in the stream. The path of the blades in vertical axis turbines is perpendicular to the direction of the flow only in two points of their circular trajectory; hence, their functioning has little correspondence with the concept of "actuator disk." Instead, horizontal axis turbines are more similar to the concept of "actuator disk of infinitesimal thickness" postulated in Betz's mathematical model, hence their real functioning is more similar to the ideal one.

b. The higher complexity of "exotic" turbines, represented in Figure 2.15 by the classic model *Gyromill®*, does not necessarily translate into higher aerodynamic efficiency. In general, neither the "inventors" nor the manufacturers of "super-Betz" wind-driven machines have ever been able (or have no interest) to produce the curve of C_p as a function of λ to demonstrate their claims. Some manufacturers of wind turbines equipped with concentrator or diffusor declare efficiencies close to, or even higher than, Betz's theoretical limit. The reason is that they only consider the area swept by the rotor, while the correct way to calculate the efficiency must consider the area of the rotor-diffusor system, because it interacts as a whole with the wind.

c. Counterintuitively, the C_p of a wind turbine is not a direct function of the number of blades. In general, increasing the number of blades means increasing the solidity, which implies that the coefficient λ must decrease (the reason will be explained in the next chapter).
   The C_p is a non-linear, but direct function of the specific speed λ.

d. From Figure 2.15 one can deduct that the theoretical C_p of Betz's ideal turbine, equal to 0.59, is an asymptotic value that can be attained only with a frictionless turbine, turning at infinite specific speed λ (and hence with solidity σ tending to zero). In reality, no existing airfoil can have a null drag coefficient, C_x; hence, the maximum values of C_p can be reached in a narrow band of specific speeds in the range of 6–10.

e. From Figure 2.15 we can deduct that even big turbines, which are always equipped with sophisticated control systems, are not able to reach C_p values greater than 0.45.

2.6 Accessory Devices of Wind Turbines

Vertical axis turbines, because of their own nature of *pannemones*, in general lack accessory devices, such as rudders and pitch controls. On the contrary, horizontal axis turbines need different accessory systems, which are not always present when the turbine is small. Examples of such accessory devices are rudders and orientation systems for the rotor, control systems of the speed and/or of the power output of the alternator, and protection devices against extreme winds. All the said accessories will be treated in detail in Chapter 4.

2.7 Exercises

2.7.1 Application of Betz's Theorem

2.6.1.1. A company claims that its "innovative" vertical axis wind turbine is capable of producing 1 kW in the wind of 10 m/s. The diameter of the rotor is 1 m and its height is 1.6 m. Check if such claim is feasible.

Solution
According to Betz's formula:

$$P_{max} = \frac{8}{27} \cdot \rho \cdot S \cdot V_1^3 = \frac{8}{27} \cdot \left[1.23 \, \text{kg} / \text{m}^3 \cdot 1.6 \, \text{m}^2 \cdot 10^3 \, \text{m}^3 / \text{s}^3 \right]$$

$$P_{max} = 583 \, \text{W}$$

The theoretically extractable energy is much smaller than the output power claimed by the manufacturer, hence such claim is false.

2.6.1.2. A company claims that one of its horizontal axis wind turbines can produce 2 kW when the wind speed is 11 m/s. The rotor's diameter is 8 (2.44 m). Check if such claim is feasible.

Solution
According to Betz's formula, the maximum theoretical power that such a turbine could extract is:

$$P_{max} = \frac{8}{27} \cdot \rho \cdot S \cdot V_1^3 = \frac{8}{27} \cdot \left[1.23 \, \text{kg} / \text{m}^3 \cdot 4.67 \, \text{m}^2 \cdot 11^3 \, \text{m}^3 / \text{s}^3 \right]$$

$$P_{max} = 2,265 \, \text{W}$$

The claimed power is too close to the theoretical limit, so most probably the company's catalogue is hyped.

2.7.2 Application of Dimensionless Coefficients

2.7.2.1. Calculate the C_P of the turbine in the former exercise 2.7.1.2

Solution
By definition:

$$C_P = \frac{2 \cdot P}{\rho \cdot S \cdot V^3} = \frac{2 \cdot 2,265 \, \text{W}}{1.23 \cdot 4.67 \cdot 11^3} = 0.5925$$

Betz's theorem establishes that the C_P of an ideal turbine is 0.59259, so the catalogue's claim cannot be true. Compare the calculated value to the best C_P of real wind turbines shown in Figure 2.16.

2.7.2.2. An international cooperation project requires a design for a water pumping system for a rural area. The design wind speed is 4 m/s. The required power is 500 W. Calculate:

a. The diameter and rotation speed in the case of a classical multi-blade windmill.
b. The diameter of a windmill with the same number of blades and rotation speed, when the blades are built with suitable airfoils instead of curved steel plates. Assume that under such condition, $C_P = 0.38$.
c. The wind load on the turbine's hub in both cases.
d. The nominal torque in both cases.
e. The start torque when $V_1 = 2.5$ m/s.

Solutions

a. Classical windmills have a maximum C_P in the order of 0.3 when their specific speed is $\lambda = 1$ (Figure 2.16). The application of the definitions is straightforward:

$$C_P = \frac{2 \cdot P}{\rho \cdot S \cdot V^3} = \frac{2 \cdot 500}{1.23 \cdot S \cdot 4^3} = 0.3$$

$$S = 42 \text{ m}^2 \rightarrow D = 7.3 \text{ m}$$

$$\lambda = \frac{\omega \cdot R}{V} = \frac{\pi \cdot D \cdot N}{60 \cdot V} \rightarrow N = \frac{60 \cdot 4}{\pi \cdot 7.3} = 10.5 \text{ R.P.M.}$$

$$C_P = \frac{2 \cdot P}{\rho \cdot S \cdot V^3} = \frac{2 \cdot 500}{1.23 \cdot S \cdot 4^3} = 0.38$$

b. $S = 33.16 \text{ m}^2 \rightarrow D = 6.5 \text{ m}$

$$\lambda = \frac{\omega \cdot R}{V} = \frac{\pi \cdot D \cdot N}{60 \cdot V} \rightarrow N = \frac{60 \cdot 4}{\pi \cdot 6.5} = 11.7 \text{ R.P.M.}$$

c. Since we demonstrated that $C_P = C_F$, the calculation of F is straightforward too.

$$F = \frac{\rho}{2} \cdot C_F \cdot S \cdot V^2$$

$$F_1 = \frac{1.23}{2} \cdot 0.3 \cdot 42 \cdot 4^2 = 123.9 \text{ N}$$

$$F_2 = \frac{1.23}{2} \cdot 0.38 \cdot 33 \cdot 4^2 = 123.4 \text{ N}$$

d. Since we demonstrated that

$$C_P = C_M \cdot \lambda$$

and $\lambda = 1$ in both cases, the calculation of M is straightforward

$$M = \frac{\rho}{2} \cdot C_M \cdot R \cdot S \cdot V^2$$

$$M_1 = \frac{1.23}{2} \cdot 0.3 \cdot 3.65 \cdot 42 \cdot 4^2 = 452 \text{ Nm}$$

$$M_2 = \frac{1.23}{2} \cdot 0.38 \cdot 3.25 \cdot 33.16 \cdot 4^2 = 403 \text{ Nm}$$

e. According to Figure 2.13, when the relative speed of a multi-blade turbine is null, $C_M \approx 0.45$, a value obtained experimentally in a wind tunnel. Unfortunately, we do not have an experimental C_M curve for a multi-blade rotor built with profiled blades. We can compare the aerodynamic data of a cambered plate to those of a thin cambered profile (see Chapter 12). The lift of the latter is 7%–10% higher. Hence, the start torque of each turbine model will be:

$$M_1 = \frac{1.23}{2} \cdot 0.45 \cdot 3.65 \cdot 42 \cdot 2.5^2 = 265 \text{ Nm}$$

$$M_1 = \frac{1.23}{2} \cdot 0.45 \cdot 3.25 \cdot 33.16 \cdot 2.5^2 = 186 \text{ Nm}$$

Bibliography

Bertin, J., *Aerodynamics for Engineers*, 4th edition, Prentice Hall, New Jersey, 2002.

Bunlung, N., Somporn, S., and Somchai, C., *Control Strategies for Variable-speed Fixed-pitch Wind Turbines*, Wind Power, INTECH, Croatia, 2010, ISBN 978-953-7619-81-7.

Cádiz Deleito, J.C. and Ramos Cabreros, J., *La Energía Eólica: Tecnología e Historia*, Hermann Blume Ediciones, Madrid, 1984.

Le Gourière, D., *L'Énergie Éolienne*, 2nd edition, Eyrolles, Paris, 1982.

Rosato, M., *Progettazione di Impianti Minieolici*, Multimedia Course, Acca Software, Montella (AV), 2010.

Rosato, M., *Diseño de Máquinas Eólicas de Pequeña Potencia*, Editorial Progensa, Sevilla, 1992.

Selig, M., Guglielmo, J. et al., *UIUC Low-Speed Airfoil Tests—Volume 1*, SoarTech Publications, Champaign, Illinois, 1998.

3

Simplified Aerodynamic Theory for the Design of the Rotor's Blades

3.1 Definition of the Problem

We have seen in the former chapters that the air, flowing through a wind turbine in movement, loses speed in the flow direction. Because wind turbines usually lack directional vanes or deflectors, the rotational movement of the blades generates a tangential component of the air motion. Simultaneously, the expansion of the flow generates a radial component. Experiments demonstrate that such radial component is negligible, while the tangential component is not, because it produces vortexes that cause energy losses.

Several mathematical models exist that explain the functioning of rotors. In order to design a small wind turbine, we can employ a complete but simplified theory that has been validated experimentally by several authors. The basic design assumptions will be the following:

a. Infinitely upstream of the turbine, the air moves along the x-axis with speed V_1, and the pressure is P_0.

b. Infinitely downstream of the turbine, the air moves along the x-axis with speed $V_2 < V_1$, and furthermore has a rotational motion whose angular speed is ω. The atmospheric pressure is always P_0.

c. An infinitesimal distance upstream of the rotation plane of the turbine, the pressure is P, the speed along the x-axis is V and furthermore the rotation of the blades with angular speed Ω induce on the air a rotational motion with the same angular speed.

d. An infinitesimal distance downstream of the rotation plane of the turbine, the pressure is P_1, the speed along the x-axis is still V, but the rotational motion induced on the air has an angular speed equal to $(\Omega + \omega)$.

e. We will analyze the flow through a differential annular area, dS, placed at a generic distance r from the axis of rotation of the turbine. The thickness of such infinitesimal annular area is dr.

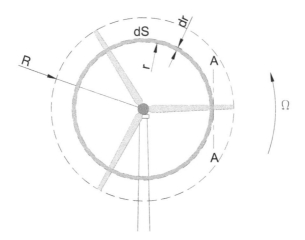

FIGURE 3.1
Front view of the turbine and of the differential ring area in study.

Figure 3.1 represents the turbine seen from the upstream side, and the corresponding magnitudes. Please, observe that:

$$dS = 2 \cdot \pi \cdot r \cdot dr$$

Figure 3.2 represents the turbine seen from the side, assumed as a disk of infinitesimal thickness. The dotted line represents the path of an air particle: the motion is unidirectional upstream of the rotor and vortical downstream of the same. The angular speeds Ω, ω, and $(\Omega+\omega)$ are represented by their vectors (orthogonal to the rotation plane, and hence parallel to the x-axis) and in red color. Because of the action and reaction principle, the rotational

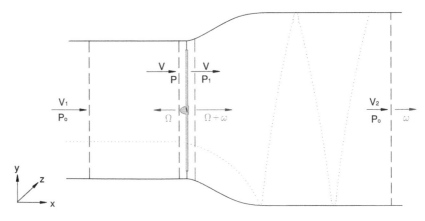

FIGURE 3.2
Lateral view of the turbine, of the speeds, pressures, and flow through a differential element of the rotor.

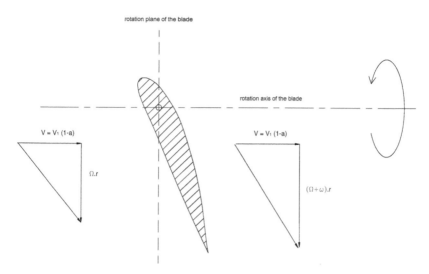

FIGURE 3.3
Vectorial composition of the speeds of a generic element of a blade of the rotor in movement (corresponding to segment A-A shown in Figure 3.1).

motion of the air particles has the opposite sense of the blades' motion. The linear speeds V_1, V, and V_2 are represented by the vectors in black color.

Looking at Figure 3.3, we can observe the composition of the speeds upstream and downstream of the element of blade marked with the segment A-A in Figure 3.1.

In order to calculate the power produced by a differential element of blade, we will first employ the equations of the quantity of motion, or momentum, from the classical Newtonian mechanics and, in a second stage, the aerodynamic forces applied to an element of the blade, already described in Chapter 2. The integration extended to the whole length of the blade will provide an expression to calculate the total power. Knowing the maximum power obtainable from a blade, we will then be able to define, in each point of the same blade, the optimal chord and pitch angle of the desired airfoil. Before attempting to calculate the power in the presence of vortexes and the derivative magnitudes, it is necessary to define some auxiliary coefficients that simplify the calculations.

3.1.1 Speed Loss Coefficient, *a*

Since the speed through the rotor, V, is not known *a priori*, it becomes useful to define a non-dimensional coefficient that allows you to express V as a function of the wind's speed upstream of the blade, V_1. The coefficient *a* is hence defined as:

$$a = 1 - \frac{V}{V_1} \quad \text{hence} \quad V = V_1 \cdot (1 - a)$$

While demonstrating Betz's theorem (Section 2.1) we found the following relationship between V, V_1, and V_2:

$$V = \frac{V_1 + V_2}{2}$$

Introducing the equivalent expression of V as a function of V_1, we obtain:

$$V = V_1 \cdot (1 - a) = \frac{V_1 + V_2}{2}$$

$$2 \cdot V_1 \cdot (1 - a) - V_1 = V_2$$

$$V_2 = V_1 \cdot (2 - 2a - 1)$$

$$V_2 = V_1 \cdot (1 - 2a)$$

Hence, the coefficient a allows us to calculate some speeds as a function of the others, at our convenience. Even if the value a is not known *a priori*, we can furthermore deduce that the same must be in the range 0–1/2, and we should ideally procure to bring it as close as possible to 1/3. Such conclusion is a consequence of Betz's theorem because, under the ideal conditions of maximum power, the value of a must be equal to 1/3. A value of a equal to 1/2 would turn V_2 to null, which is physically impossible. On the other side, if a was null, then V_2 would be equal to V_1, hence the turbine would not turn.

3.1.2 Coefficient of Specific Local Speed, λ_r

It is the coefficient λ defined at a generic point of the blade, placed at a distance r from the center.

Remember that λ was defined for the blade's tip as:

$$\lambda = \frac{u}{V_1} = \frac{\Omega \cdot R}{V_1}$$

In the same way, the coefficient λ_r at a generic point will be:

$$\lambda_r = \frac{\Omega \cdot r}{V_1} = \frac{\lambda \cdot r}{R}$$

3.1.3 Coefficient of Angular Speed, a'

The angular speed of the turbine is equal to Ω. The interaction of the rotor's blades with the air induce on this latter a rotational component, whose angular speed is equal to ω. The ratio between both angular speeds is defined as:

$$a' = \frac{\omega}{2 \cdot \Omega}$$

3.2 The Theory of the Annular Flow Tube with Vortical Trail

The following reasoning will help us to find the relationships linking a, a', and λ. Knowing the said relationships, we will then be able to calculate the values of V, V_2, and ω. Hence, knowing the relative speeds at each point of the blade, it will be possible to apply the aerodynamic theory to calculate the forces on a differential section, and by integration, on the whole blade. Our first goal will be to find the conditions that maximize the power in order to design the geometry of the blade for optimal performances.

Observing Figures 3.1 and 3.2, we can apply the equation of energy conservation in a flow (a.k.a. *Bernoulli's* equation) to a section placed at an infinite distance upstream of the turbine and to another section placed at an infinitesimal distance upstream of the turbine. Under such conditions, the Bernoulli's equation becomes:

$$P_0 + \frac{\rho \cdot V_1^2}{2} = P + \frac{\rho \cdot V^2}{2}$$

If we now apply Bernoulli's equation to a section placed an infinitesimal distance downstream of the turbine and to another section placed at infinite distance downstream of the turbine, we will obtain the following equation:

$$P_0 + \frac{\rho \cdot V_2^2}{2} = P_1 + \frac{\rho \cdot V^2}{2}$$

Subtracting member from member equations, we get:

$$P_0 + \frac{\rho \cdot V_1^2}{2} - P_0 - \frac{\rho \cdot V_2^2}{2} - = P + \frac{\rho \cdot V^2}{2} - P_1 - \frac{\rho \cdot V^2}{2}$$

$$\frac{\rho \cdot V_1^2}{2} - \frac{\rho \cdot V_2^2}{2} = P - P_1 = \Delta P$$

Recalling the relationships between V_1, V_2 and the coefficient a postulated in Section 3.1.1, we can replace in the former equation the speed V_2 (unknown) with its equivalent as a function of V_1 and a.

$$\Delta P = \frac{\rho}{2} \cdot (V_1^2 - V_2^2) = \frac{\rho}{2} \cdot \left[V_1^2 - V_1^2 \cdot (1-2a)^2\right]$$

$$\Delta P = \frac{\rho \cdot V_1^2}{2} \cdot \left[1 - (1-2a)^2\right] = \frac{\rho \cdot V_1^2}{2} \cdot (1 - 1 + 4a - 4a^2)$$

$$\Delta P = 2 \cdot \rho \cdot V_1^2 \cdot a \cdot (1-a)$$

The axial force applied by the wind on the differential annular crown dS can be calculated as follows:

$$dF_a = \Delta P \cdot dS = 2 \cdot \rho \cdot V_1^2 \cdot a \cdot (1-a) \cdot 2 \cdot \pi \cdot r \cdot dr$$
$$dF_a = 4 \cdot \pi \cdot \rho \cdot V_1^2 \cdot a \cdot (1-a) \cdot r \cdot dr \tag{3.1}$$

We will now apply Bernoulli's equation to both sides of the circular crown, dS. From Figure 3.3 it is evident that, since the flow was assumed as bidimensional (components of the air speed only in x and in y), we must consider the vectorial resultants of the said components of speed, V_m and V_v (the suffixes indicate, respectively, the resultants upstream and downstream of the area dS).

We obtain:

$$P_1 + \frac{\rho \cdot V_m^2}{2} = P_2 + \frac{\rho \cdot V_v^2}{2}$$

By Pythagoras' theorem we get:

$$V_m^2 = V^2 + (\Omega \cdot r)^2 = V_1^2 \cdot (1-a)^2 + (\Omega \cdot r)^2$$
$$V_v^2 = V^2 + (\Omega + \omega)^2 \cdot r^2 = V_1^2 \cdot (1-a)^2 + (\Omega + \omega)^2 \cdot r^2$$

Replacing the values of V_m and V_v in Bernoulli's equation and reordering the terms, we get:

$$P_1 - P_2 = \Delta P = \frac{\rho}{2} \cdot \left[V^2 + (\Omega + \omega)^2 \cdot r^2 - V^2 - (\Omega \cdot r)^2 \right]$$

$$\Delta P = \frac{\rho \cdot r^2}{2} \cdot \left[(\Omega + \omega)^2 - \Omega^2 \right]$$

$$\Delta P = \frac{\rho \cdot r^2}{2} \cdot \left[(\Omega^2 + 2\Omega\omega + \omega^2) - \Omega^2 \right]$$

$$\Delta P = \frac{\rho \cdot r^2}{2} \cdot \left[(2\Omega\omega + \omega^2) \right]$$

Introducing the coefficient a' defined in Section 3.1.3, we can now replace the unknown component, ω, by its equivalent as a function of the known component, Ω:

$$\Delta P = \frac{\rho \cdot r^2}{2} \cdot \left[2\Omega \cdot (2\Omega \cdot a') + (2\Omega \cdot a')^2 \right]$$

$$\Delta P = \frac{\rho \cdot r^2}{2} \cdot \left[4\Omega^2 \cdot a' + 4\Omega^2 \cdot a'^2 \right]$$

$$\Delta P = \frac{\rho \cdot r^2 \cdot 4 \cdot \Omega^2}{2} \cdot \left[a' + a'^2 \right]$$

$$\Delta P = 2 \cdot \rho \cdot r^2 \cdot \Omega^2 \cdot a' \cdot \left[1 + a' \right]$$

The axial thrust on the annular crown dS will then be:

$$dF_a = \Delta P \cdot dS = 2 \cdot \rho \cdot r^2 \cdot \Omega^2 \cdot a' \cdot (1 + a') \cdot (\pi \cdot 2r \cdot dr)$$

$$dF_a = 4 \cdot \pi \cdot \rho \cdot \Omega^2 \cdot a' \cdot (1 + a') \cdot r^3 \cdot dr \tag{3.2}$$

So, we have obtained two equivalent expressions for the force dF_a, one as a function of the coefficient a and of the speed V_1, and the other as a function of the coefficient a' and of the angular speed Ω.

Equalling Equation 3.1 with Equation 3.2 we obtain:

$$4 \cdot \pi \cdot \rho \cdot V_1^2 \cdot a \cdot (1 - a) \cdot r \cdot dr = 4 \cdot \pi \cdot \rho \cdot \Omega^2 \cdot a' \cdot (1 + a') \cdot r^3 \cdot dr$$

$$V_1^2 \cdot a \cdot (1 - a) = \Omega^2 \cdot a' \cdot (1 + a') \cdot r^2$$

$$\frac{a \cdot (1 - a)}{a' \cdot (1 + a')} = \frac{\Omega^2 \cdot r^2}{V_1^2}$$

Remembering the definition of the coefficient of local specific speed, λ_r, presented in Section 3.1.2, the former equation becomes:

$$\lambda_r^2 = \frac{a \cdot (1 - a)}{a' \cdot (1 + a')} \tag{3.3}$$

Now that we know the relationships between a, a', and λ_r, we can calculate the power produced by the differential section of blade. From Figure 3.3, we deduce that, since the axial component of the wind speed, V, remains constant at both sides of the blade, only the tangential component will produce work. In order to calculate the tangential force on the element of the blade, please remember the principle of impulse and momentum (Newton's second law):

$$F \cdot t = m \cdot \Delta V$$

where F is the force acting on a portion of matter having mass m during a certain time t, and ΔV indicates the variation of the speed that the mass m will then experience. We can then deduce that:

$$F = \frac{m}{t} \cdot \Delta V = q \cdot \Delta V$$

because, by definition, the quotient between a unit of mass and a unit of time is the flow, q. Assuming that the air is incompressible, which is fairly true under the normal operational conditions of a wind turbine, the flow dq across the area dS shown in Figure 3.1 can be calculated as:

$$dq = \rho \cdot dS \cdot V$$

From Figure 3.3 we deduce that, when the air flows through the turbine's rotor, the component of axial flow does not experience speed variations (assumption of Betz's theorem), while the tangential flow component will undergo a speed variation, ΔV, given by the following equation:

$$\Delta V = (\Omega + \omega) \cdot r - \Omega \cdot r = \omega \cdot r$$

Consequently, we can express the tangential force of the air acting on the differential element of blade inside the area dS as:

$$dF_t = \rho \cdot dS \cdot V \cdot \omega \cdot r$$

We can then replace dS by its complete expression, and V by its equivalent as a function of the wind's speed, V_1, and of the coefficient a:

$$dF_t = \rho \cdot (2 \cdot \pi \cdot r \cdot dr) \cdot [V_1 \cdot (1 - a)] \cdot \omega \cdot r$$

$$dF_t = 2 \cdot \pi \cdot \rho \cdot V_1 \cdot (1 - a) \cdot \omega \cdot r^2 \cdot dr$$

Now we replace the unknown magnitude ω by its equivalent, using the coefficient a':

$$dF_t = 2 \cdot \pi \cdot \rho \cdot V_1 \cdot (1 - a) \cdot 2 \cdot \Omega \cdot a' \cdot r^2 \cdot dr$$

$$dF_t = 4 \cdot \pi \cdot \rho \cdot V_1 \cdot (1 - a) \cdot \Omega \cdot a' \cdot r^2 \cdot dr \tag{3.4}$$

The power produced by such force, by definition, is the product of the torque at the center of rotation by the angular speed, hence we get:

$$dP = 4 \cdot \pi \cdot \rho \cdot V_1 \cdot (1 - a) \cdot \Omega \cdot a' \cdot r^2 \cdot dr \cdot r \cdot \Omega$$

$$dP = 4 \cdot \pi \cdot \rho \cdot V_1 \cdot (1 - a) \cdot \Omega^2 \cdot a' \cdot r^3 \cdot dr$$

Replacing Ω^2 by its equivalent as a function of λ_r.

$$dP = 4 \cdot \pi \cdot \rho \cdot V_1^3 \cdot (1 - a) \cdot \frac{\lambda_r^2}{r^2} \cdot a' \cdot r^3 \cdot dr$$

$$dP = 4 \cdot \pi \cdot \rho \cdot V_1^3 \cdot (1 - a) \cdot \lambda_r^2 \cdot a' \cdot r \cdot dr \tag{3.5}$$

Since $\lambda_r > 0$ in all cases, the power will be maximum when the product $a' \cdot (1 - a)$ will be maximum, i.e., when its derivative is null. Since a and a' are functions of each other, then the derivative of the product of two functions (in our case the function $(1-a)$ and the function a') is calculated with the rule of the product:

$$y = f_{(x)} \cdot g_{(x)} \Rightarrow y' = f'_{(x)} \cdot g_{(x)} + f_{(x)} \cdot g'_{(x)}$$

Hence:

$$\frac{d}{da}(1-a) \cdot a' + (1-a)\frac{da'}{da} = 0$$

$$(1-a) \cdot \frac{da'}{da} = a'$$

$$(1-a) \cdot da' = a' \cdot da \tag{3.6}$$

We know that a and a' are linked by the Equation 3.3, from which we can write:

$$\lambda_r^2 \cdot a' \cdot (1+a') = a \cdot (1-a)$$

Since a' is a function of a, we can differentiate each member of the equation in respect to a, applying the "rule of the chain":

$$D[f_{(g(x))}] = f'_{(g(x))} \cdot g'_{(x)}$$

We will then obtain:

$$\lambda_r^2 \cdot (1+2a') \cdot \frac{da'}{da} = (1-2a)$$

$$\lambda_r^2 \cdot (1+2a') \cdot da' = (1-2a) \cdot da \tag{3.7}$$

Multiplying member to member Equations 3.6 and 3.7, we obtain:

$$\lambda_r^2 \cdot (1+2a') \cdot a' \cdot da \cdot da' = (1-2a) \cdot (1-a) \cdot da \cdot da'$$

$$\lambda_r^2 \cdot (1+2a') \cdot a' = (1-2a) \cdot (1-a)$$

Replacing now the value of λ_r^2 from Equation 3.3, we get:

$$\frac{a \cdot (1-a)}{a' \cdot (1+a')} = \frac{(1-2a) \cdot (1-a)}{(1+2a') \cdot a'}$$

$$\frac{a}{(1+a')} = \frac{(1-2a)}{(1+2a')}$$

$$a + 2aa' = 1 - 2a + a' - 2aa'$$

$$4aa' - a' = 1 - 3a$$

$$a' = \frac{1-3a}{4a-1} \tag{3.8}$$

Summarizing, Equations 3.3 and 3.8 link a, a', and λ, allowing to maximize the power, as will be shown below.

In Equation 3.5, we can now replace r by its equivalent as a function of λ_r, already defined in Section 3.1.2.

$$dP = 4 \cdot \pi \cdot \rho \cdot V_1^3 \cdot a' \cdot (1-a) \cdot \lambda_r^2 \cdot \frac{R \cdot \lambda_r}{\lambda} \cdot dr$$

Always based on the same equivalence defined in Section 3.1.2, we can replace dr as follows:

$$\frac{dr}{d\lambda_r} = \frac{R}{\lambda} \Rightarrow dr = \frac{R}{\lambda} \cdot d\lambda_r$$

$$dP = 4 \cdot \pi \cdot \rho \cdot V_1^3 \cdot a' \cdot (1-a) \cdot \lambda_r^2 \cdot \frac{R \cdot \lambda_r}{\lambda} \cdot \frac{R}{\lambda} \cdot d\lambda_r$$

$$dP = 4 \cdot \pi \cdot \rho \cdot V_1^3 \cdot \frac{R^2}{\lambda^2} \cdot a' \cdot (1-a) \cdot \lambda_r^3 \cdot d\lambda_r$$

Hence, it is possible to obtain the total power produced by the turbine by integrating between $\lambda_r = 0$ (center of the rotor) and $\lambda_r = \lambda$ (tip of the blade). Consequently:

$$P = 4 \cdot \pi \cdot \rho \cdot V_1^3 \cdot \frac{R^2}{\lambda^2} \int_0^\lambda a' \cdot (1-a) \cdot \lambda_r^3 \cdot d\lambda_r$$

Under such condition, the coefficient of power C_P is given by:

$$C_P = \frac{P}{\dfrac{\rho}{2} V_1^3 \cdot \pi \cdot R^2} = \frac{4 \cdot \pi \cdot \rho \cdot V_1^3 \cdot R^2}{\dfrac{\rho}{2} V_1^3 \cdot \pi \cdot R^2 \cdot \lambda^2} \int_0^\lambda a' \cdot (1-a) \cdot \lambda_r^3 \cdot d\lambda_r$$

$$C_P = \frac{8}{\lambda^2} \int_0^\lambda a' \cdot (1-a) \cdot \lambda_r^3 \cdot d\lambda_r, \text{ known as } Glauert's \ integral \tag{3.9}$$

The algebraic resolution of *Glauert's integral* is extremely difficult, if not impossible, because of the complexity of the relationships between a, a', and λ_r. In 1935, the English mathematician and aerodynamic researcher *Herman Glauert* solved such equation with a graphic method. Today, it is much easier to tabulate a' and λ_r as a function of a with the help of a spreadsheet, and then to calculate by discrete integration the maximum value of C_P resulting for each value of λ. Our task is made easier because the value of a is limited to the intervals 1/2 and 1/3, otherwise a has no physical sense. Hence, it is possible to divide the said interval in n discrete intervals and to integrate by simply adding all the adjacent cells. Table 3.1 is an example of the calculation in the interval of λ_r between 0 (center of rotation of the blade) and $\lambda_r = 0.5$.

Figure 3.4 shows the resulting C_P, within the interval $0 < \lambda < 12$.

The theory of Glauert, condensed in Figure 3.4, highlights that *the aerodynamic efficiency grows with λ and tends asymptotically to Betz's limit*. Glauert's

TABLE 3.1

Example of Calculation of Formulas 3.3, 3.8, and the Integral in Equation 3.9, with the Help of a Spreadsheet

a	λ_r	a'	Integral	C_P
0.25	0	0	0	0
0.251	0.006966	61.75	1.09E-07	0.017952
0.252	0.014007	30.5	5.5E-07	0.022441
0.253	0.021127	20.0833333	1.56E-06	0.027917
0.254	0.028327	14.875	3.37E-06	0.033635
0.255	0.03561	11.75	6.25E-06	0.039446
0.256	0.042979	9.66666667	1.05E-05	0.045301
0.257	0.050435	8.17857143	1.63E-05	0.051179
0.258	0.057983	7.0625	2.4E-05	0.057068
0.259	0.065623	6.19444444	3.39E-05	0.062965
0.26	0.073361	5.5	4.63E-05	0.068865
0.261	0.081198	4.93181818	6.16E-05	0.074767
0.262	0.089138	4.45833333	8.01E-05	0.08067
0.263	0.097184	4.05769231	0.000102	0.086573
0.264	0.10534	3.71428571	0.000128	0.092476
0.265	0.11361	3.41666667	0.000159	0.098379
0.266	0.121998	3.15625	0.000194	0.104281
0.267	0.130507	2.92647059	0.000235	0.110183
0.268	0.139143	2.72222222	0.000281	0.116086
0.269	0.147909	2.53947368	0.000334	0.121989
0.27	0.15681	2.375	0.000393	0.127893
0.271	0.165853	2.22619048	0.00046	0.133798
0.272	0.175041	2.09090909	0.000535	0.139705
0.273	0.184381	1.9673913	0.000619	0.145615
0.274	0.193878	1.85416667	0.000712	0.151528
0.275	0.20354	1.75	0.000815	0.157444
0.276	0.213372	1.65384615	0.00093	0.163364
0.277	0.223383	1.56481481	0.001056	0.16929
0.278	0.233578	1.48214286	0.001195	0.175221
0.279	0.243967	1.40517241	0.001348	0.181159
0.28	0.254558	1.33333333	0.001516	0.187103

(Continued)

TABLE 3.1 (*Continued*)

Example of Calculation of Formulas 3.3, 3.8, and the Integral in Equation 3.9, with the Help of a Spreadsheet

a	λ_r	a'	Integral	C_P
0.281	0.265361	1.26612903	0.001699	0.193056
0.282	0.276383	1.203125	0.0019	0.199018
0.283	0.287637	1.14393939	0.00212	0.20499
0.284	0.299133	1.08823529	0.00236	0.210972
0.285	0.310883	1.03571429	0.002621	0.216966
0.286	0.3229	0.98611111	0.002906	0.222974
0.287	0.335196	0.93918919	0.003216	0.228995
0.288	0.347788	0.89473684	0.003554	0.235032
0.289	0.36069	0.8525641	0.003921	0.241086
0.29	0.373919	0.8125	0.00432	0.247157
0.291	0.387494	0.77439024	0.004753	0.253248
0.292	0.401435	0.73809524	0.005224	0.25936
0.293	0.415763	0.70348837	0.005737	0.265494
0.294	0.430501	0.67045455	0.006293	0.271653
0.295	0.445675	0.63888889	0.006898	0.277838
0.296	0.461312	0.60869565	0.007556	0.284051
0.297	0.477444	0.57978723	0.008272	0.290295
0.298	0.494102	0.55208333	0.00905	0.296571
0.298347	0.500004	0.54275128	0.009331	0.298601

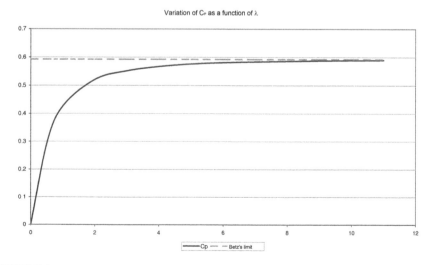

Variation of C$_P$ as a function of λ.

FIGURE 3.4
Variation of the coefficient of power C_P, as a function of λ, of a real blade with negligible drag, compared to Betz's ideal limit for an "actuator disk." Re-elaboration of *Glauert*'s integral (1935) performed by the Author with a spreadsheet.

theory in itself is not enough to design a wind turbine's blade, but the expressions obtained above are useful to define the border conditions that each element of the blade must satisfy, so that the latter results optimum as a whole.

3.3 The Theory of the Aerodynamic Forces on the Element of Blade

In this section, we are going to face the design of the blades from the point of view of the aerodynamic forces acting on the differential element of blade, already defined in Figures 3.1 and 3.3. We will employ the results obtained in Section 3.2 by means of Glauert's theory of the induced speeds (in particular, the expressions of a and a').

It is important to make a short digression for a caveat before proceeding with the exposition of the calculation method: the design of big commercial turbines generally employs calculation models, which are more sophisticated than the one we will explain in the following paragraphs. Since the blades of small-sized turbines operate in low-Re conditions and, considering that the profile of the wind speeds near ground is irregular, it has little sense to base the design of the blades on more complex models. The validity of the method described in this section has been checked in the wind tunnel by several authors, which are mentioned in the Bibliography.

As a start point, please observe Figure 3.1. The experience teaches us that an element of the blade comprising between r and $r+dr$, and immersed in a fluid stream, will be subject to a system of forces. The forces depend on the shape of the blade's profile, and on the pitch angle between the chord of the profile and the relative speed of the air, W. Figure 3.5 shows the relative speed between the blade and the air, derived from Figure 3.3 and from Glauert's theory.

The angles shown in Figure 3.5 are defined as follows:

α = angle of incidence of the flow, at a blade section defined at a generic distance r from the center of rotation

φ = pitch angle of the section of the profile, a.k.a. local pitch of the blade section at a distance r from the center of rotation

θ = angle between the rotation plane of the blade and the relative speed W, at a generic distance r from the center of rotation

From the same figure it is possible to deduce that:

$$\theta = \alpha + \varphi$$

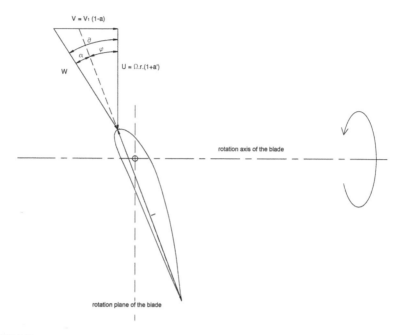

FIGURE 3.5
Sketch of the relative speeds on a differential element of blade placed at a generic distance r from the center of rotation.

Furthermore:

$$tg\theta = \frac{V_1 \cdot (1-a)}{\Omega \cdot r \cdot (1+a')} = \frac{(1-a)}{\lambda \cdot (1+a')}$$

(3.10)

From Figure 3.5, we can furthermore deduce that:

$$W^2 = U^2 + V^2 = V_1^2 \cdot (1-a)^2 + \Omega^2 \cdot r^2 \cdot (1+a')^2$$

The relative speed W, incident on the profile with angle α, generates the following aerodynamic forces:

$$dF_z = \text{lift}$$

$$dF_x = \text{drag}$$

The values of these forces can be calculated with the following expressions:

$$dF_z = \frac{\rho}{2} C_z \cdot l \cdot W^2 \cdot dr$$

$$dF_x = \frac{\rho}{2} C_x \cdot l \cdot W^2 \cdot dr$$

where l is the length of the chord of the profile, for the moment unknown. Our goal is to determine the optimum value of l for a given pair of values C_z and C_x, chosen from the airfoil's polar or table of aerodynamic data.

Figure 3.6 shows the forces F_z and F_x, respectively, perpendicular and coaxial to the relative speed W, and their axial and tangential projections: F_f (a force that generates only a flexion load on the blade, but does not produce useful work), and F_u (a force that generates useful work). When referring to the differential element of blade, we will call these forces, respectively, dF_f and dF_u.

Applying the fundamental trigonometric relationships, we can express the value of dF_u as:

$$dF_u = dF_z \cdot \mathrm{sen}\theta - dF_x \cdot \cos\theta$$

We can then describe dF_u as:

$$dF_u = \frac{\rho}{2} C_z \cdot l \cdot W^2 \cdot dr \cdot \mathrm{sen}\theta - \frac{\rho}{2} C_x \cdot l \cdot W^2 \cdot dr \cdot \cos\theta \tag{3.11}$$

With the same reasoning, we can write:

$$dF_f = \frac{\rho}{2} C_z \cdot l \cdot W^2 \cdot dr \cdot \cos\theta + \frac{\rho}{2} C_x \cdot l \cdot W^2 \cdot dr \cdot \mathrm{sen}\theta \tag{3.12}$$

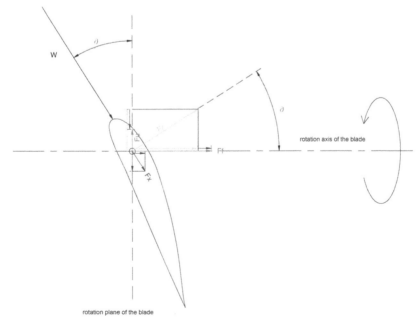

FIGURE 3.6
Aerodynamic forces and their axial and tangential projections.

In an airfoil operating under optimum conditions, the relationship $C_z \gg C_x$ must be satisfied always, and so we can simplify the Equations 3.11 and 3.12 as follows:

$$dF_u \approx \frac{\rho}{2} C_z \cdot l \cdot W^2 \cdot dr \cdot \text{sen}\,\theta \qquad (3.13)$$

$$dF_f \approx \frac{\rho}{2} C_z \cdot l \cdot W^2 \cdot dr \cdot \cos\theta \qquad (3.14)$$

If the number of blades is equal to z, we can calculate the axial force dF_a acting on the differential annular area as follows:

$$dF_a = \frac{\rho}{2} \cdot z \cdot C_z \cdot l \cdot W^2 \cdot dr \cdot \cos\theta$$

Replacing W^2 with its equivalent as a function of θ, we obtain:

$$dF_a = \frac{\rho}{2} \cdot z \cdot C_z \cdot l \cdot \frac{V_1^2 \cdot (1-a)^2}{\text{sen}^2\theta} \cdot dr \cdot \cos\theta \qquad (3.15)$$

Comparing Equation 3.12 with Equation 3.1, obtained with the theory of the induced speeds, we get:

$$dF_a = 4 \cdot \pi \cdot \rho \cdot V_1^2 \cdot a \cdot (1-a) \cdot r \cdot dr = \frac{\rho}{2} \cdot z \cdot C_z \cdot l \cdot \frac{V_1^2 \cdot (1-a)^2}{\text{sen}^2\theta} \cdot dr \cdot \cos\theta$$

$$4 \cdot a = \frac{z \cdot l}{2 \cdot \pi \cdot r} \cdot C_z \cdot \frac{(1-a)}{\text{sen}^2\theta} \cdot \cos\theta$$

In the same way, we have defined the coefficient of solidity of a turbine, σ, (Section 2.3.3.1), we will now define the coefficient of solidity for a differential ring of the same turbine. This new coefficient is called *local solidity*, σ_l, and is defined as:

$$\sigma_l = \frac{z \cdot l}{2 \cdot \pi \cdot r}$$

Hence, we can rearrange the equations of dF_a as follows:

$$\frac{a}{(1-a)} = \frac{\sigma_l \cdot C_z \cdot \cos\theta}{4 \cdot \text{sen}^2\theta} \qquad (3.16)$$

The useful force dF_u generates a torque with respect to the axis of rotation. Choosing arbitrarily the number of blades (z), we can then calculate the resulting differential total torque:

$$dM = z \cdot r \cdot dF_u \qquad (3.17)$$

From Equation 3.4 we can deduce that:

$$dM = dF_t \cdot r = 4 \cdot \pi \cdot \rho \cdot V_1 \cdot (1-a) \cdot \Omega \cdot a' \cdot r^3 \cdot dr$$

Equalling the last expression to Equation 3.17, we get:

$$4 \cdot \pi \cdot \rho \cdot V_1 \cdot (1-a) \cdot \Omega \cdot a' \cdot r^3 \cdot dr = z \cdot r \cdot dF_u$$

Replacing the expression of dF_u from Equation 3.13 into the last equation above, the result is:

$$4 \cdot \pi \cdot \rho \cdot V_1 \cdot (1-a) \cdot \Omega \cdot a' \cdot r^3 \cdot dr = z \cdot r \cdot C_z \cdot \frac{\rho}{2} \cdot l \cdot W^2 \cdot dr \cdot sen\theta$$

which we can simplify and rearrange in the following manner:

$$4 \cdot V_1 \cdot (1-a) \cdot \Omega \cdot a' \cdot r = C_z \cdot \frac{z \cdot l}{2 \cdot \pi \cdot r} \cdot W^2 \cdot sen\theta$$

Replacing W^2 by its trigonometric equivalent (from Figure 3.5) and introducing the coefficient σ_l, the equation above remains:

$$4 \cdot V_1 \cdot (1-a) \cdot \Omega \cdot a' \cdot r = \sigma_l \cdot C_z \cdot \frac{V_1^2 \cdot (1-a)^2}{sen^2\theta} \cdot sen\theta$$

$$4 \cdot \Omega \cdot a' \cdot r = \sigma_l \cdot C_z \cdot \frac{V \cdot (1-a)}{sen\theta}$$

$$a' = \sigma_l \cdot C_z \cdot \frac{V \cdot (1-a)}{4 \cdot \Omega \cdot r \cdot sen\theta}$$

Replacing $V \cdot (1-a)$ with its trigonometric equivalent from Equation 3.10, we can express a' as follows:

$$a' = \sigma_l \cdot C_z \cdot \frac{\Omega \cdot r \cdot (1+a') \cdot tg\theta}{4 \cdot \Omega \cdot r \cdot sen\theta}$$

$$\frac{a'}{1+a'} = \sigma_l \cdot C_z \cdot \frac{tg\theta}{4 \cdot sen\theta}$$

$$\frac{a'}{1+a'} = \frac{\sigma_l \cdot C_z}{4 \cdot cos\theta} \tag{3.18}$$

Replacing a' in Equation 3.18 by its equivalent from Equation 3.8, we obtain:

$$\frac{\dfrac{1-3a}{4a-1}}{1+\dfrac{1-3a}{4a-1}} = \frac{\sigma_l \cdot C_z}{4 \cdot cos\theta}$$

$$\frac{1-3a}{4a-1+1-3a} = \frac{\sigma_l \cdot C_z}{4 \cdot \cos\theta}$$

$$\frac{1-3a}{a} = \frac{\sigma_l \cdot C_z}{4 \cdot \cos\theta}$$

$$\frac{1}{a} = \frac{\sigma_l \cdot C_z}{4 \cdot \cos\theta} + 3$$

$$a = \frac{4 \cdot \cos\theta}{\sigma_l \cdot C_z + 12 \cdot \cos\theta} \qquad (3.19)$$

Replacing Equation 3.19 into Equation 3.16, we get

$$\frac{\dfrac{4 \cdot \cos\theta}{\sigma_l \cdot C_z + 12 \cdot \cos\theta}}{\left(1 - \dfrac{4 \cdot \cos\theta}{\sigma_l \cdot C_z + 12 \cdot \cos\theta}\right)} = \frac{\sigma_l \cdot C_z \cdot \cos\theta}{4 \cdot \mathrm{sen}^2\theta}$$

$$\frac{\dfrac{4 \cdot \cos\theta}{\sigma_l \cdot C_z + 12 \cdot \cos\theta}}{\dfrac{\sigma_l \cdot C_z + 12 \cdot \cos\theta - 4 \cdot \cos\theta}{\sigma_l \cdot C_z + 12 \cdot \cos\theta}} = \frac{\sigma_l \cdot C_z \cdot \cos\theta}{4 \cdot \mathrm{sen}^2\theta}$$

$$\frac{4}{\sigma_l \cdot C_z + 8 \cdot \cos\theta} = \frac{\sigma_l \cdot C_z}{4 \cdot \mathrm{sen}^2\theta}$$

$$4 = \frac{(\sigma_l \cdot C_z)^2 + \sigma_l \cdot C_z \cdot 8 \cdot \cos\theta}{4 \cdot \mathrm{sen}^2\theta}$$

$$16 \cdot \mathrm{sen}^2\theta = (\sigma_l \cdot C_z)^2 + \sigma_l \cdot C_z \cdot 8 \cdot \cos\theta$$

Please note that this latter is a second-degree equation, in which the variable is the product $\sigma_l \cdot C_z$. The two roots are:

$$\sigma_l \cdot C_z = 4 \cdot (-1 - \cos\theta) \text{ and}$$

$$\sigma_l \cdot C_z = 4 \cdot (1 - \cos\theta) \qquad (3.20)$$

The first has no physical sense because $\cos\theta > 0$ in the interval of validity, in order that $\sigma_l \cdot C_z > 0$, hence only the Equation 3.20 is valid. We have then obtained an equation that links directly the local solidity, σ_l, and the chord of the profile, l, with the angle θ.

The coefficient C_z is arbitrarily chosen by the designer on the base of given performance criteria, and the value of C_x results as a consequence, because their relationship is univocal. For instance, it is possible to choose the optimum lift value, i.e., the C_z corresponding to the angle at which the ratio C_z/C_x is maximum (maximum aerodynamic efficiency of the profile). It is necessary to consider that the values of C_z and C_x published in the aerodynamic literature, are conventionally defined for a wing of infinite length. In a real wing, and in particular at the tip of a turbine's blade, the difference of pressure between the intrados and the extrados of the airfoil will produce a vortex, because there is no border condition preventing its formation. It is common to observe commercial aircraft having "winglets" or plates at the tips of their wings, which are important to reduce the drag and to improve the efficiency in fuel consumption. Such technique has been applied rarely to wind turbines. The reason is that the aspect ratio of wind turbine blades is usually much bigger than that of aircraft wings, so placing a plate or winglet at the blade's tip brings little benefit. From the classical aerodynamic theory, it is known that the drag caused by tip vortexes can be calculated with Oswald's Formula:

$$C_{xi} = \frac{C_z^2}{\pi \cdot e \cdot AR}$$

where:

C_{xi} = drag coefficient induced by the vortexes

C_z = lift coefficient (from aerodynamic tables, valid for wings of infinite length)

e = Oswald's empirical coefficient = 0.85–0.9 for very elongated blades or wings

AR = aspect ratio = $L/l = L^2/A$

L = total length of the blade (measured from the connection to the hub up to the tip)

l = average chord of the blade or wing

A = area of the blade = $L \cdot l$

In practice, we must employ in our calculations a corrected drag coefficient, C_{xr}, which is the sum of the value of C_x, obtained from aerodynamic tables, and the induced drag from Oswald's formula, which is a function of C_z:

$$C_{xr} = C_x + C_{xi}$$

At this point, we finally have all the necessary elements to design each section of the blade. The following section will present some additional considerations that will turn useful for the design of a horizontal axis wind turbine.

3.3.1 Optimum Variation of the Angle θ

It is desirable to extract the maximum of energy from the wind at each position r of the blade; consequently, a and a' must satisfy Equations 3.3 and 3.8. Since the angle θ results from Equation 3.10, we deduce that θ is a function of λ_r. In the same way that we have tabulated the local coefficient of power as a function of a, a', and λ_r, we can calculate the optimum angle θ at each position r along the blade, because λ_r is directly proportional to r,

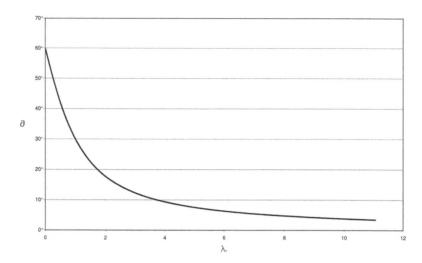

FIGURE 3.7
Optimal variation of the angle θ as a function of λ_r.

according to the definition presented in Section 3.1.2. Figure 3.7 shows how θ varies as a function of λ_r.

The curve shown in Figure 3.7 can be represented with the following equivalent mathematical function, which is easier to employ in a spreadsheet than Equation 3.10:

$$\theta = \frac{2}{3}\operatorname{arctg}\left(\frac{1}{\lambda_r}\right) \tag{3.21}$$

Since, according to Equation 3.20, the product $\sigma_l \cdot C_z$ depends only on cos θ, once its optimum value is known for each value of λ_r, and once the values of R and of λ are defined, we can calculate the product $\sigma_l \cdot C_z$ directly with the Equation 3.20, replacing Equation 3.21 in it.

For instance, suppose we define a constant value of the chord l along the blade, and we calculate σ_l for each position of r, applying the definition of the local solidity coefficient:

$$\sigma_l = \frac{z \cdot l}{2 \cdot \pi \cdot r}$$

Now that σ_l is known, we can calculate the value of C_z that satisfies Equation 3.20, in each point of the blade. Having chosen C_z from the aerodynamic tables of the airfoil, we know the attack angle α and hence we are able to define the pitch of the section of the blade at the point r, i.e., the angle φ of the Figure 3.5. From a practical point of view, such approach is useful, for instance, to produce a blade from an extruded profile: since the chord is constant along the blade, the manufacturer can deform the extruded profile by applying torsion to it, in order that the angle φ in each point will satisfy the following equations:

$$\theta = \alpha + \varphi; \theta = \frac{2}{3}\text{arctg}\left(\frac{1}{\lambda_r}\right); \alpha = \alpha' - \beta \text{ and } \beta = \text{arctg}\left(\frac{C_z \cdot \pi \cdot l}{R}\right)$$

A blade designed with such criterion is sub-optimum, because the resulting values of C_z and the corresponding incidence angle, in general are not coincident with the point of maximum aerodynamic efficiency of the profile. If the hypothesis on which Equations 3.13 and 3.14 was postulated ($C_z \gg C_x$) is not satisfied, the mathematic model explained in Section 3.3 yields less accurate results.

3.3.2 Optimum Variation of the Product $\sigma_l \cdot C_z$

As shown in Equation 3.20, the product $\sigma_l \cdot C_z$ depends only on cos θ and hence, if the optimum value of θ is known for each value of λ_r, and having already predefined the design values R and λ, we can calculate the product $\sigma_l \cdot C_z$ directly with the Equation 3.20. For instance, if we desire to build a blade from a simple wooden or plywood plank (hence φ = constant), we need to define arbitrarily a constant value for the angle φ in Figure 3.5. Having defined the angle φ at each point r of the blade, calculating the angle θ with Equation 3.10, it is immediate to determine the resulting angle of incidence, α. Once α is known, the corresponding C_z results from the airfoil's data table, and since the product $\sigma_l \cdot C_z$ is known, the calculation of the chord l in each section of the blade is straightforward.

3.3.3 Optimum Blade for Maximum Aerodynamic Efficiency

The lift to drag ratio or fineness coefficient, $f = C_z/C_x$, represents the aerodynamic efficiency of an airfoil. In the former sections, we have based all our design assumptions on the fact that $C_z \gg C_x$, which is true only at small values of the angle of incidence (in general less than 5°). The ratio f is maximum only in a point of the airfoil's polar. We can design the blades of a turbine in order to maximize the energy extracted from the wind, if in each point r the Equations 3.10 and 3.20 are simultaneously satisfied, and furthermore, if the value of C_z corresponding to the maximum f is chosen. Thus, the hypothesis of Equations 3.13 and 3.14 are fulfilled in each section of the blade. The resulting shape of the blade will feature a certain twist and its chord will be variable from the hub to the tip. Since λ_r is directly proportional to r, we can tabulate the product $\sigma_l \cdot C_z \cdot \lambda_r$ as a function of λ_r, which in turn is a function of a and a', in the same way employed to plot the curve shown in Figure 3.7. The plot of said function represents, in a dimensionless mode, the variation of the chord along the blade. The result is shown in Figure 3.8.

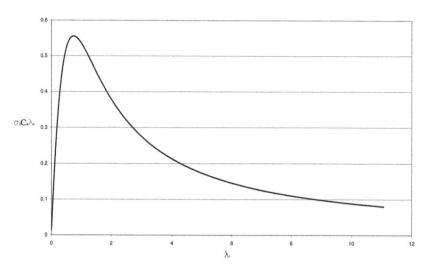

FIGURE 3.8
Variation of the optimum chord, for C_z = constant.

3.4 Conclusions

From the treatise in Section 3.3, and from the digressions in Sections 3.3.1 and 3.3.3, it is possible to draw the following conclusions:

3.4.1 Variation of the Chord

When designing a slow turbine, in which λ is comprised between 0 and 1, the chord is always growing from the hub to the tip.

When designing a fast turbine, in which $\lambda > 2$, the chord grows from the hub until the point where $\lambda \approx 1$ and then decreases from the said point until the tip.

3.4.2 Relationship between Solidity, Specific Speed, and Efficiency of the Turbine

Recalling the concept of solidity, defined as the quotient between the area of the blades and the total swept area of the rotor, and expressing it mathematically in general form, we obtain:

$$\sigma = \int_0^R \frac{z \cdot l}{\pi \cdot R^2} dr$$

Since:

$$\lambda_r = \frac{\Omega \cdot r}{V_1} \Rightarrow r = \frac{V_1 \cdot \lambda_r}{\Omega} \Rightarrow \frac{dr}{d\lambda_r} = \frac{V_1}{\Omega} \Rightarrow dr = \frac{V_1}{\Omega} d\lambda_r$$

Replacing dr in the integral, we get:

$$\sigma = \int_0^\lambda \frac{z \cdot l}{\pi \cdot R^2} \frac{V_1}{\Omega} d\lambda_r$$

It is possible to multiply denominator and numerator by some factors, in order to group variables and constants, without altering the equation, as follows:

$$\sigma = \int_0^\lambda \frac{z \cdot l}{\pi \cdot R^2} \frac{V_1}{\Omega} \frac{\Omega \cdot C_z \cdot V_1 \cdot 2\pi}{\Omega \cdot C_z \cdot V_1 \cdot 2\pi} d\lambda_r$$

Grouping the constants and bringing them out of the integral sign, we finally obtain:

$$\sigma = \frac{2\pi \cdot V_1^2}{C_z \cdot \pi \cdot \Omega^2 \cdot R^2} \int_0^\lambda \frac{z \cdot l \cdot \Omega \cdot C_z}{V_1 \cdot 2\pi} d\lambda_r$$

Please note that the terms inside the integral correspond to the product $\sigma_l \cdot C_z \cdot \lambda_r$, and those outside of the integral can be further grouped and simplified as follows:

$$\sigma = \frac{2}{C_z \cdot \lambda^2} \int_0^\lambda \sigma_l \cdot C_z \cdot \lambda_r \cdot d\lambda_r$$

It is possible to solve numerically this integral with a spreadsheet, assuming, for example, $C_z = 0.8$ (NACA 4412 airfoil with $\alpha = 5°$). The result is shown in Figure 3.9.

3.4.2.1 Solidity and Specific Speed

Figure 3.9 tells us that the solidity of the turbine, σ, decreases with increasing specific speed λ.

From Figure 3.9 it is furthermore evident that $\sigma > 1$ when $\lambda < 1$. Such result is apparently contradictory, but it is not physically impossible. The reason is that, in general, slow turbines have many blades that can eventually overlap, hence the sum of the blades' area can turn to be bigger than the area swept by the rotor. In practice, it is impossible that the blade starts from the

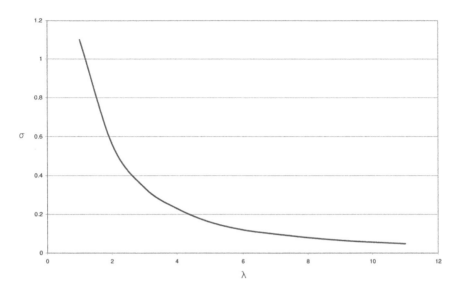

FIGURE 3.9
Solidity of the turbine as a function of its specific speed assuming $C_z = 0.8$ as a constant along the blade.

very geometric center, which is occupied by the hub. Slow turbines are those featuring $\lambda \leq 1$.

3.4.2.2 Solidity and Aerodynamic Efficiency

From *Glauert's Theory*, whose result is graphically summarized in Figure 3.5, it is possible to deduce that the theoretical coefficient of power, C_P, grows with λ. Since the solidity decreases with λ, wind turbines featuring low solidity (and hence a small number of blades) are always more efficient than those having high solidity and a large number of blades.

3.4.3 Influence of the Fineness Coefficient of the Airfoil

When explaining the theory of the element of blade, we neglected the aerodynamic drag of the airfoil, with the scope of simplifying the formulas to obtain Equations 3.13 and 3.14. Such assumption is reasonably valid for slender airfoils, perfectly smooth, and operating with Re > 500,000. Under the real operational conditions of a wind turbine (profiles thick enough to grant the structural resistance, dirty surface because of dust and dead insects, surface scratched by flying sand, Re < 100,000 in case of very little turbines and/or very weak winds) it is seldom true that $C_z \gg C_x$ or, conversely, that $f > 50$. Figure 3.10 shows how the coefficient of power C_p decreases with the fineness coefficient f.

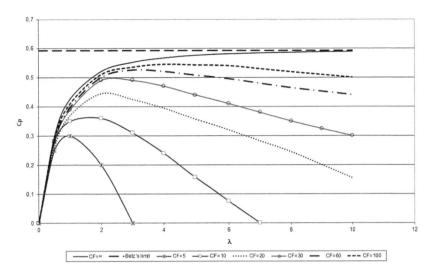

FIGURE 3.10
Comparison of the C_p as a function of λ, for ideal turbines without induced drag, and for different finite values of f. Data from *Le Gourière*, graphic elaboration by the Author.

3.5 Practical Exercises

3.5.1 Influence of the Induced Drag

Compare the induced drag of the blades in three different wind turbines, all of them having 3 m diameter. Assume Re = 200,000 and blades' chord and pitch angle designed to be optimum. Calculate the resulting coefficient f and check the maximum theoretical C_p from Figure 3.10.

3.5.1.1 Classical Windmill for Water Pumping

The same has 20 blades made with curved steel plates, not overlapping. Assume the aerodynamic features of a cambered plate to be those of the GOE 417A airfoil (Section 12.3.5). The total solidity is $\sigma = 0.95$.

Solution
Since the diameter is 3 m and the solidity is 0.95, the blades start at a radius r that must satisfy the following condition:

$$\sigma = 0.95 = \frac{\pi \cdot (R^2 - r^2)}{\pi \cdot R^2}$$

$$r = 1.5\,\text{m}\sqrt{(1 - 0.95)} = 0.33\,\text{m}$$

The length of each blade is then:

$$L = 1.5 - 0.33 = 1.17\,\text{m}$$

and its AR is:

$$AR = \frac{1.17^2}{\left[(0.95 \cdot \pi \cdot 1.5^2)/20\right]} = 4$$

At its best performance point ($\alpha = 5°$), the curved plate features $C_z = 1.014$ and $C_x = 0.014$. The induced drag is then:

$$C_{xi} = \frac{C_z^2}{\pi \cdot e \cdot AR} = \frac{1.014^2}{\pi \cdot 0.9 \cdot 4} = 0.09$$

and the total drag is:

$$C_{xr} = C_x + C_{xi} = 0.014 + 0.09 = 0.104$$

The effective lift to drag ratio is then: $f = 1.014/0.104 = 9.7$

3.5.1.2 Multi-Blade Turbine

Suppose building a "bicycle wheel turbine", i.e., a turbine featuring a very high number of very thin blades. Assume our model will have 120 blades. Assume the latter are made with cambered plates having the same aerodynamic features of airfoil GOE 417A.

Solution
Repeating the same procedure already shown in the former example, we get: $AR = 24.5$; $C_{xi} = 0.015$; $C_{xr} = 0.029$. The effective lift to drag ratio is then: $f = 1.014/0.029 = 35$

3.5.1.3 A Three-Blade Fast Turbine

Assume that the airfoil employed to design the blades is NREL S822 (Section 12.3.4). The solidity is $\sigma = 0.05$ and the blades have the same length L as in the former case.

Solution

$$AR = \frac{1.17^2}{\left[(0.05 \cdot \pi \cdot 1.5^2)/3\right]} = 11.6$$

At its best performance point ($\alpha = 7°$), the airfoil features $C_z = 0.987$ and $C_x = 0.014$. The induced drag is then:

$$C_{xi} = \frac{C_z^2}{\pi \cdot e \cdot AR} = \frac{0.987^2}{\pi \cdot 0.9 \cdot 11.6} = 0.03$$

and the total drag is:

$$C_{xr} = C_x + C_{xi} = 0.014 + 0.03 = 0.044$$

The effective lift to drag ratio is then: $f = 0.987/0.044 = 22.4$

3.5.1.4 Practical Conclusions

The aerodynamic performance of curved plates at low Re and infinite wing-span is slightly better than that of an airfoil specially designed for fast wind turbines, like the NREL series, at the same ideal conditions. Nevertheless, the high solidity of traditional windmill rotors implies that the aspect ratio of their single blades is very low, resulting in a high value of induced drag, bringing the blade's total drag to nearly 10 times the drag of the cambered plate with infinite span. On the contrary, the blades of fast wind turbines usually feature $AR > 10$. Hence, their induced drag is nearly three times that of the infinite span airfoil, and the resulting total drag is nearly four times the same. A rotor with a high number of very thin blades, like the one supposed in Section 3.5.1.2, has potentially a high aerodynamic performance at lowRe, but it presents several drawbacks:

- heavy construction and consequently higher material cost;
- higher mechanical loads in strong winds—and consequently more cost for reinforced structures;
- the shape of the blades is optimized for $\lambda = 1$, but the best theoretical C_P can be attained in the interval $3 < \lambda < 4$.

Bibliography

Burton T., Sharpe D., Jenkins N., and Bossanyi E., *Wind Energy Handbook*, John Wiley & Sons Inc., Chichester, 2001.

Hansen M., *Aerodynamics of Wind Turbines*, 2nd edition, Earthscan, London, 2008.

Le Gourière D., *L'Énergie Éolienne – Théorie, conception et calcul practique des installations*, 2nd edition, Eyrolles, Paris, 1982.

Manwell J.F., Mcgowan J.G., and Rogers A.L., *Wind Energy Explained: Theory, Design and Application*, 2nd edition, John Wiley & Sons Inc., Chichester, 2010.

Rosato M., *Diseño de máquinas eólicas de pequeña potencia*, Editorial Progensa, Sevilla, 1992.

Rosato M., *Progettazione di Impianti Minieolici*, Multimedia Course, Acca Software, 2010.

4

Practical Design of Horizontal
Axis Wind Turbines

4.1 Generalities

We will analyze in depth the phases and details of a wind turbine's design in the next chapters. For the moment, the general procedure can be summarized as follows:

1. Defining the turbine's scope: Electrical generation or water pumping.
2. Analysis of the wind potential of the site intended for the installation of the turbine. This argument is so vast, that it would require a dedicated book to treat it in depth. In Chapter 8, we will present some of the classical statistical functions employed in the wind power industry. The practical examples at the end of the same chapter describe some techniques of simulation with spreadsheets, based on tabular anemometric data.
3. Pre-dimensioning and analysis of the rotor's aerodynamic features in order to obtain the desired performances.
4. Final verification and dimensioning.
5. Calculation of the mechanical resistance of the turbine's elements. This aspect will be treated in Chapter 7.
6. Study of the efficiency in converting the wind energy into the desired energy form, and its potential storage technology. This is the argument of Chapters 9 and 10, for the electric and water storage, respectively.

In the present chapter, we will discuss the general design approach of horizontal axis wind turbine rotors, regardless of the same being slow or fast.

4.2 The Method to Design the Rotor

The input data to design a wind turbine rotor are: the desired kind of turbine (slow for water pumping, fast for electrical generation), the required nominal power, and the nominal wind speed, which is a characteristic of the site. The next sections explain step by step the calculation sequence.

4.2.1 Pre-Dimensioning of the Diameter and Number of Blades

Assuming that the characteristic wind speed of the site is known, and that the type of turbine has already been decided, we will calculate the diameter of the rotor and its rotational speed with the formulas already presented in Chapter 2, Section 2.2.4.1.

4.2.1.1 Pre-Dimensioning Fast Turbines

The diameter results from Betz's formula applied to real turbines. If we assume $C_P \approx 0.4$ and $\rho \approx 1.21$, the formula reduces to:

$$P = 0.20 \cdot D^2 \cdot V_1^3 \rightarrow D = \sqrt{\frac{P}{0.20 \cdot V_1^3}}$$

Stating P in W and V_1 in m/s in the former equation, the diameter of the rotor results in m.

The number of blades, z, can be 1, 2, or 3 in fast turbines. Single-blade turbines present some issues of vibrations and difficulty of dynamic balancing, hence they are not practical. The designer must then choose between fast rotors with 2 or 3 blades. The 2-bladed rotor is easier to build, but the majority of manufacturers prefer the 3-bladed solution because of its higher stability of functioning.

In general, fast turbines are suitable for $\lambda \geq 5$ and $V > 5$ m/s. *Le Gourière* advises to define the value of λ as a function of the lift to drag coefficient, f, of the chosen airfoil. As a rule of thumb, we can assume the indicative values shown in Table 4.1.

Now that V_1, D, and λ are known, the nominal rotation speed, N, remains automatically defined because of its univocal relationship with λ (Chapter 2, Section 2.3.3.2), hence:

$$\lambda = \frac{\Omega \cdot R}{V_1} = \frac{\pi \cdot D \cdot N}{60 \cdot V_1} \rightarrow N = \frac{\lambda \cdot 60 \cdot V_1}{\pi \cdot D}$$

where:
  Ω = Angular speed (rad/s)
  N = Angular speed (R.P.M.)

TABLE 4.1

Empirical Relationship Between the Airfoil's Fineness
Coefficient, f, the Number of Blades, z, and the Specific Speed, λ

f	λ	z
5	0.5–1	12–24
20–40	2–3	8–12
40–50	4–5	6–8
60–100	5–7	4–6
>100	7–13	2 or 3

N.B.: The values of f considered in this table include the induced drag
and the influence of Re.

V_1 = Nominal speed of the wind, characteristic of the site (m/s)
$R = D/2$ = Radius (m)

4.2.1.2 Pre-Dimensioning of Slow Turbines

As in the former case, the diameter results from Betz's formula generalized
for real turbines, assuming $C_P \approx 0.3$ and $\rho \approx 1.21$. The formula reduces to the
following expression:

$$P = 0.15 \cdot D^2 \cdot V_1^3 \rightarrow D = \sqrt{\frac{P}{0.15 \cdot V_1^3}}$$

In the former equation, P is expressed in W, and V_1 in m/s, hence the diameter of the rotor results in m.

In order to predefine the specific speed λ and the number of blades z, we
can employ the empirical data published by *Le Gourière*, shown in Table 4.1.

Defining arbitrarily the value of λ, and knowing V_1, we will obtain the
rotation speed N (R.P.M.) or Ω (rad/s) with the formulas already explained
in Section 4.2.1.1.

$$N = \frac{\lambda \cdot 60 \cdot V_1}{\pi \cdot D}$$

4.2.2 Dimensioning of the Yaw System: Vane or Rotor Conicity

Horizontal axis turbines always need a yaw system to keep the rotor oriented perpendicularly to the direction of the wind. There are three possible
yaw systems for small power turbines:

a. Orientation by means of a vane.
b. Orientation by means of conicity.
c. Orientation by means of a servomotor (only in bigger turbines, typically 20 kW or more).

4.2.2.1 Orientation by Means of a Vane

This is probably the most common among small wind turbines and the area of the vane, s, can be calculated by means of the following empirical formulas, recommended by *Le Gourière*:

$$\text{For fast turbines: } s = 0.16 \cdot \frac{\pi \cdot D^2}{4} \cdot \frac{E}{L}$$

$$\text{For slow turbines: } s = 0.4 \cdot \frac{\pi \cdot D^2}{4} \cdot \frac{E}{L}$$

Figure 4.1 shows the meaning of each factor in the formulas above. The shape of the vane is irrelevant and in general, it becomes a kind of distinctive sign of each manufacturer.

4.2.2.2 Orientation by Conicity

Self-orientable conical rotors are easier to build than the ones formerly described, but may present some issues because of vibrations. These are caused by the path of the blade across the "wind shadow cone" produced by

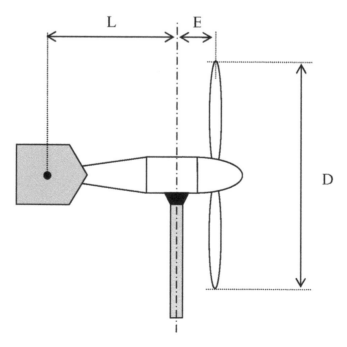

FIGURE 4.1
Reference dimensions for the design of the vane.

the tower or support structure, which generates vortexes. The intensity and frequency of the vortexes is a function of the diameter of the support pole.

Tests on a "big" prototype turbine (500 kW) with conical rotor, built by the National Aeronautics and Space Administration (NASA) in the 1970s, were interrupted because of the very bothersome vibrations at low frequency it caused.

According to *Le Gourière*, who quotes experiences conducted by prof. *Cambilargiu*, of the University of Montevideo, the vibrations can be minimized by keeping the rotation plane of the blades at a certain distance x from the support pole. Such distance must be comprised between 25% and 30% of the diameter of the rotor, D, on the condition that the diameter of the pole, d, is equal to 2.2% of the diameter of the rotor D (see Figure 4.2).

4.2.2.3 Orientation by Means of a Servomotor

Yaw control by servomotor requires sophisticated devices and is of course more expensive than the solutions presented above. In general, such devices

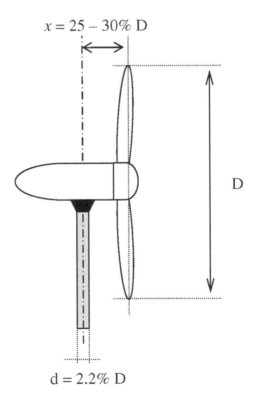

$x = 25 - 30\%$ D

D

d = 2.2% D

FIGURE 4.2
Sketch of the empirical dimensions recommended for self-orientable conical rotor turbines, in order to minimize vibrations.

are suitable for turbines bigger than 20 kW. There are some commercial exceptions, for instance, an Italian model called TN 7®, which has 6.5 kW nominal power.

The principle of functioning of analogic controls is the following: a small vane placed on top of the turbine is rigidly connected to the cursor of a potentiometer, which forms the moiety of a Wheatstone's resistive bridge. The body of the potentiometer, in turn, is aligned to the hub of the turbine. When the turbine is correctly oriented, the cursor coincides with the center of the potentiometer, the bridge is then in equilibrium, hence the servomotor's driver receives no input signal and the servomotor does not turn, leaving the turbine faced to the wind. If the wind changes direction, the cursor will move away from the center, bringing the bridge out of equilibrium. The voltage that now appears across the bridge, i.e., the signal to the servomotor's driver, is now different from zero, so the driver feeds the servomotor, which in turn will move the turbine until recovering the equilibrium position. Digital yaw controls are based on a device called shaft encoder, and a microprocessor that calculates the angular difference between the vane's shaft encoder and that of the turbine's yaw mechanism.

In any case, servo-assisted yaw controls always have a complementary anemometer that, in case of excessive wind speed, will generate an additional signal to the driver which in turn will cause the Wheatstone's bridge to reach the equilibrium in a different position, precisely when the rotor's plane is parallel to the wind.

Servo-assisted yaw controls are beyond the possibilities of most small wind turbine builders, and will not be dealt further in this book.

4.2.3 Selection of the Most Suitable Airfoil for the Blades

An infinity of airfoils exists, some of them are considered "general use" while others were designed on purpose for specific scopes.

The classical text about the theory of airfoils is that of *Abbot* and *von Doehoff*, that contains a good quantity of aerodynamic data measured in the wind tunnel, most of them corresponding to National Advisory Committee for Aeronautics (NACA) airfoils. The most famous series of standard airfoils, and perhaps the oldest, is the 4-figure NACA, followed by the more modern 5- and 6-figure NACA series. NACA was the former name of today's NASA before the spatial era. The said airfoils series are hence described by the name NACA and the figures describe the basic geometric factors. For instance, in the NACA 4-figure series, the first one is the maximum value of the camber line (expressed as a percentage of the chord); the second is the position of the maximum camber point (in tenths of the chord, measured from the leading edge); the third and fourth figures represent the maximum thickness of the airfoil (expressed as a percentage of the chord's length). For example, a symmetric airfoil having thickness equal to 12% of the chord will be called NACA 0012 (because, having no camber at all, the first two figures are null).

An airfoil with maximum camber equal to 4% of the chord's length, l, placed at a point $0.4l$ from the leading edge, and thickness 12% of l, will be then called NACA 4412.

More modern airfoils, developed by today's NASA, do not follow the same nomenclature: an airfoil developed during the 1980s is called LS(1), and a variant of the same is LS(1) MOD. Many other series of airfoils exist, developed by either individual researchers or universities. Typically, all of them are called as the developer (or the name's abbreviation), followed by a number. For instance, airfoil DU-93-W-210, which is employed in some big turbines manufactured in Europe, was developed at the University of Delft; airfoil E 385 was developed by a researcher called Eppler; airfoil RAF 6 was developed by the British Royal Air Force during the World War I.

Until the 1980s, the selection of the airfoil used to be a marginal aspect in the design of a small wind turbine's rotor. Typically the designer's choice used to fall on 4-figure NACA airfoils because empirical data of helicopter propellers, made with airfoils NACA 0012 and NACA 4412, were available. The comparison between the functioning in windmill and propeller modes somehow served to validate the mathematical models for the calculation of the turbines, and to have an estimation of the noise. Such comparison is not fully valid, because wind turbines run with much lower Re than aeronautical propellers, and furthermore are installed near to the ground, where the air is always "dirtier" than that at the flight heights of aircraft. Placing the turbines at low heights implies that the leading edge of the blades will get "aerodynamically rugose" because of insects colliding with the turning rotor. Furthermore, sand and coarse dust transported by the wind will scratch the blades' surface. The increase of the rugosity caused by such environmental factors leads to a drastic drop of the airfoil's aerodynamic efficiency (defined as $f = C_z/C_x$). The so-called laminar airfoils of the NACA 5- and 6-figure series, have better stall behavior and high f at low pitch angles, but are anyway sensitive to the leading edge's dirtiness.

For example, an airfoil NACA 63-415 having its surface perfectly smooth and operating at $Re = 3 \cdot 10^6$, has a maximum fineness coefficient $f = 120$ when the pitch angle is 4°. The same airfoil, at the same operational conditions, but having "NACA standard rugosity," has $f = 67$.

"NACA standard rugosity" is a treatment applied to models tested in the wind tunnel, consisting of gluing sand with an average particle diameter equal to 0.0011 (27.7 μm), to the extrados and intrados between the leading edge and 8% of the airfoil's chord, where the standard chord for wind tunnel tests is 24 (605 mm). The sand is sprayed in such a way that it covers from 5% to 10% of the area defined before.

More modern airfoils, like NASA LS(1) MOD, tolerate better the rugosity, but their aerodynamic performances can drop to 40% of the initial f value in very dusty zones as the blades are subject to erosion, or dead insects accumulate on their surfaces.

More recently, some airfoil families special for wind turbines have been developed, as for instance National Renewable Energy Laboratory (NREL) and Eppler.

In 1980, the researchers *Eppler* and *Somers* developed a model of calculation that was able to generate shapes of airfoils compliant with predefined specifications. The researchers of the NREL, based on Eppler and Somers's program, developed the NREL S-Series of airfoils, which are able to tolerate the increase of rugosity without suffering drastic drops of their fineness coefficient. Table 4.2 summarizes the results from *NREL Airfoil Families for HAWTs*, by J.L. Tangler and D.M. Somers, about the increase of annual productivity that the NREL airfoils can provide, compared to the traditional NACA series.

The "airfoils families" consist of three or four geometrically similar airfoils, having different thickness. The thickest is useful for making the blade's root, because of structural reasons, while the thinnest is ideal for the blade's tip, where reaching the maximum aerodynamic efficiency is the designer's main goal.

The blades designed in the said way are very complex to build, because the thickness varies along the radius, and consequently, the aerodynamic features of each section will be different. The lift and drag in the intermediate sections must be estimated by interpolation. In general, small turbines ($D < 10$ m) employ the same airfoil from the blade's root to the tip, because it is easier to calculate and build such a rotor. The compromise between good aerodynamic performances and structural resistance in small wind turbines consists of employing airfoils with thickness comprised between 10% and 15% of the chord.

From the point of view of the small turbines builder, very often the primary criterion of airfoil selection is the manufacturing simplicity. In this sense, the simplest airfoils for hand-made rotors are cambered metal or plastic plates and flat-convex airfoils. The first type of airfoils has been employed for over a century for the manufacture of multi-blade turbines for water pumping. The second type of airfoils is represented by the *Clark Y* profile, which is easy to build with plywood and basic hand tools.

Chapter 12 includes data of some selected airfoils, taken from different sources. The author has normalized the graphic display of the airfoils geometry, and provided all data as tables, because it is easier to integrate them in the calculation models.

TABLE 4.2

Increase of Productivity of the NREL Airfoils Compared to Similar NACA Airfoils

Type of Turbine	Overall Efficiency Increase (%)
Self-regulated by stall	23–35
Variable pitch, constant speed	8–20
Fixed pitch, variable speed	8–10

Researchers and aircraft model amateurs employ "virtual wind tunnels," calculation programs that are able to simulate the aerodynamic behavior of any streamlined geometry defined arbitrarily. A large collection of data, freely accessible on the Internet, can be found in the following sites www. airfoiltools.com and http://www.airfoildb.com/.

Table 4.3 summarizes the features of some airfoils that are suitable for the manufacture of small wind turbines.

TABLE 4.3

Indicative Data about Some Common Airfoils Included in Chapter 12, Which Are Useful for the Pre-selection and Comparison

Name	Possible Application	Comments
NACA 0012	Fast turbines, $\lambda > 6$	Symmetric airfoil. Lacquering and waxing the blade is advisable in order to obtain a smooth surface. High f, but sensitive to rugosity increments.
NACA 4412	Fast turbines, $\lambda > 6$	Asymmetric airfoil. Lacquering and waxing the blade is advisable in order to obtain a smooth surface. High f, but sensitive to rugosity increments.
Clark Y	Fast turbines, $\lambda = 5$–6	Flat-convex airfoil, suitable for handcrafted construction with plywood, medium-low f.
FX 77-W-153	Airfoil specially conceived for fast turbines with high aerodynamic efficiency, $\lambda = 6$–10	Difficult handcrafted construction. Tolerates some rugosity with little efficiency loss. High f.
LS(1)	NASA general purpose aviation airfoil, good performances with fast turbines, $\lambda = 6$–8	Difficult handcrafted construction. High f, medium sensitivity to rugosity increments.
Cambered plate	Slow turbines, $\lambda = 1$–3	Low f. Easy and cheap to build by cambering metallic plates, or by cutting segments along big-diameter, thin-walled plastic tubes. Little sensitivity to the variations of Re.
Eppler E220	Airfoil specially conceived for operation at low Re. Suitable for fast turbines, $\lambda = 6$–8	Difficult handcrafted construction thickness = 11.48%. High f, little sensitivity to increments of rugosity.
NREL S-822	Thick airfoil for turbines with $D < 10\,\mathrm{m}$	Tolerates the rugosity, has good structural and aerodynamic features, high f.
NREL S-819	Thick airfoil for turbines with $10\,\mathrm{m} < D < 40\,\mathrm{m}$	Tolerates the rugosity, has good structural and aerodynamic features, medium-high f.
GOE 417 A	Thin airfoil, suitable for rotors made with sails or cambered plates	Developed for the airplanes of the World War I. Mediocre aerodynamic features at low Re.
E317 MOD	Airfoil specially conceived for operation at low Re. Suitable for fast turbines, $\lambda = 6$–8	Its shape makes it particularly suitable for the construction of rotors with rigid sails, or cambered plates fitted out with a suitable leading-edge profile.

4.2.4 Division of the Blade in N Discrete "Differential Elements"

Once the length of the blade, the characteristics of the airfoil, and the other design parameters already explained are defined, it is necessary to "discretize," i.e., to define a division of the blade in elements small enough, that the chord and pitch angle can be considered constant in the chosen interval, "*dr*." The bigger the quantity of discrete differential elements, the higher the calculation accuracy. The only inconvenience is the increasing size of the spreadsheet or calculation program employed. A good compromise between size of the file and calculation accuracy is dividing the blade in 50 to 100 discrete elements. Dividing the blade in just 20 finite elements yields results with an acceptable accuracy. It is not advisable to divide the blade in less than 20 finite elements, because the "steps" of pitch angle and chord length of the airfoils between adjacent sections will make more difficult its construction.

4.2.5 Calculation of the Chord and Pitch Angle for Each Discrete Element

Three different criteria exist:

4.2.5.1 Calculation of the Optimum Chord and Pitch Angle for Each Discrete Element

Once the diameter and speed of the turbine is defined, we need to define the necessary data for the blades' design. The following parameters are either known in advance, or arbitrarily chosen by the designer: R, λ, z, r_0 (distance from the center of the rotor at which the blade will attach to the hub), N, ρ (density of air, normally assumed as 1.22–1.25 kg/m³), V and Ω.

At each point of the blade, the value λ_r becomes automatically defined and consequently, the angle θ results from Equation 3.21. Knowing θ, Equation 3.20 allows us to calculate the product $\sigma \cdot C_z$.

Now, we can arbitrarily choose the lift coefficient, C_z, and its corresponding drag coefficient, C_x, and define them to be constant along the blade. The said coefficients are taken from the characteristic curves or tables of the preselected airfoil, choosing the incidence angle that produces the maximum aerodynamic efficiency or fineness coefficient, f.

In order to account for the tip vortexes, we correct the value of C_x in each section of blade as explained in Chapter 3, Section 3.3, using Oswald's correction formula:

$$C_{xi} = \frac{C_z^2}{\pi \cdot e \cdot AR}$$

Since the desired incidence angle, α, and the angle θ are both known, the difference is equal to the local pitch angle, φ.

Since the angle θ is known, and C_z was arbitrarily chosen, we calculate the product $\sigma_l \cdot C_z$ with Equation 3.20, at each of the points in which the blade was divided:

$$\sigma_l \cdot C_z = 4 \cdot (1 - \cos\theta)$$

Since C_z is known, we are then able to calculate the optimum chord, l, by applying the definition of σ_l.

$$\sigma_l = \frac{z \cdot l}{2 \cdot \pi \cdot r}$$

4.2.5.2 Calculation of Sub-optimum Blades in Order to Facilitate the Handcrafted Construction

Another calculation method involves defining arbitrarily a law of the chord variation along the blade—usually linear variation or constant chord—and determining the optimum twist of the blade. Conversely, it is possible to arbitrarily impose a law defining the pitch angle along the blade, and then calculate accordingly the optimum chord in each point along the blade.

4.2.5.2.1 Constant or Linearly Variable Chord

Assuming a constant chord is useful in order to build the blades with extruded airfoils. Assuming a linear variation of the chord is useful to build blades with trapezoidal planform, either with plywood or metal. In both cases, the angle θ in each point of the blade is known in advance, because the product $\sigma_l \cdot C_z$ results from a predefined value of σ_l, and C_z is chosen arbitrarily. When the chord is the predefined parameter, σ_l is known in advance; hence it is necessary to calculate the value of C_z that satisfies the product. Entering the table of aerodynamic data with the calculated C_z, the corresponding values of C_x and incidence angle, α, are automatically defined. At this point, we can calculate the twist angle φ as the difference between the angles θ and α. Once the values of the chord l and C_z are known, the forces dF_u and dF_f on each segment of the blade are automatically defined.

4.2.5.2.2 Constant or Linearly Variable Twist

It is possible to build a blade with a flat plywood board, hence no twist at all, and nevertheless to obtain acceptable aerodynamic performances.

The procedure is analogous to the one already explained in the former Section 4.2.5.1. We start by defining arbitrarily a constant value of φ along the blade (for instance 10°). In other words, the plane of the blade will form an angle φ with the rotation plane of the turbine. In each point of the blade, knowing λ, and hence a, a' and θ, it is immediate to calculate the product $\sigma_l \cdot C_z$ and the angle of incidence α. Knowing the angle α, C_z, and C_x are automatically defined from the airfoil's data. Being C_z known, we can calculate

σ_l, from which we calculate the chord l. Now that the angle θ, the chord l, C_z, and C_x are known, the forces dF_u and dF_f are automatically calculable for each segment of the blade.

Regardless of the criterion chosen among the three possible options, once the chord and twist angle are defined in each point of the blade, it is possible to analyze the rotor's performance with different wind and rotation speeds, checking the resulting incidence angle α at each point of the blade, and entering as parameters the resulting values of the coefficients C_z and C_x resulting from it. This allows to plot the C_P versus λ curve and the P versus V curve.

4.2.6 Discrete Integration of the Tangential and Axial Forces along the Blade

At this point of the design process, the sum of the discrete values dF_f, calculated at each segment of the blade, multiplied by the number of blades, will give as result the total axial force F_a, useful to calculate the stress at the hub and on the support structure. The product $dF_u \cdot r$ is the differential moment of each segment, and the sum of all of them is the total motor torque. The product of the torque by the rotation speed corresponds to the power output at the shaft. Repeating the calculations for each wind and rotation speed, we can plot the curves of power output.

4.3 Analysis of the Aerodynamic Features and Construction Choices of the Rotor

Regardless of the blade being designed to be optimum or sub-optimum, now that we have defined its geometry, there are several degrees of freedom in order to design the rotor and its desired aerodynamic behavior at different wind speeds. Three possible control strategies of the rotor exist:

a. Fixed pitch and control by stall. This technique can employ either static or dynamic stall, as explained later on. In the past, some techniques of stall induced by *spoilers* were proposed too.

b. Variable pitch. This is the most diffused and sophisticated control technique in the market of big turbines. In the past, it was applied to small turbines too, but the progress in electronics tends to make the option a) simpler and cheaper.

c. Control of the exposed area or of the rotor's yaw angle. This solution is simple to build, but the regulation of the power output is imprecise. It is suitable only for very small turbines.

4.3.1 Fixed Pitch Rotor

This is the most frequent option for small turbines, because the cost of electronic inverters is affordable and hence controlling the output voltage and frequency electronically is easier than building a rotor with mechanical actuators, which are always potential failure points, and in some cases, can increase the manufacturing costs.

A fixed pitch of the blades implies that the speed of rotation will increase with growing wind speed, because the turbine tends to keep the value of λ constant, and close to the point of maximum efficiency. Such behavior is a benefit when the wind is weaker than the nominal one, a situation represented by the point A in Figure 4.3. When the wind reaches its nominal speed, the turbine runs at its maximum C_p, point B in Figure 4.3. With wind stronger than the nominal speed, C_p cannot run stably at its maximum value: the power at the shaft will be bigger than the maximum power that the generator can yield. Consequently, the rotor will increase its rotation speed. On doing so, the angle of incidence at each section of the blade will decrease, and hence the coefficient C_z will drop. The power at the rotor's shaft will drop, until it reaches an equilibrium with the power absorbed by the generator (point C in Figure 4.3). The speed of rotation will hence stabilize at a higher value than the design's nominal, and at a lower C_P.

The operation at fixed pitch and variable speed "naturally" limits the power output. Such operation can be admissible, but only to a certain extent. The designer must calculate the resulting speed of rotation for wind speeds bigger than the nominal, in order to check that safety values are not trespassed. High rotation speeds induce compound stresses on the blades, caused by

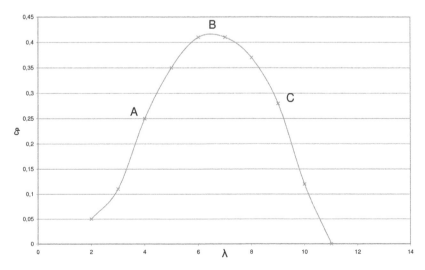

FIGURE 4.3
Functioning of a wind turbine at variable speed.

the combination of centrifugal and axial forces, and fatigue on the materials because of vibrations that arise if the rotor is not well balanced. The wind turbine engineer must assess case by case the limited speed of rotation, depending on the availability of tools and instruments for the dynamic balancing of the rotor, its dimensions, and the materials employed for its manufacture.

A common design criterion consists in limiting the blade tip's speed to 80 m/s, not only for structural but also for acoustical reasons. It becomes then necessary to foresee some kind of braking system which can be centrifugal, aerodynamic, electronic, mechanical, or combination of the same. The next section presents the strategies of speed control for fixed pitch turbines.

4.3.1.1 Fixed Speed Rotor and Unlimited (or Very High Limit) Speed of Rotation

This kind of rotor is suitable for turbines having very small diameter, indicatively smaller than 2 m, that can be balanced dynamically in the workshop. By employing symmetric airfoils (e.g., NACA 0012), when the wind speed grows, the rotation speed grows proportionally, and λ grows. A growing value of λ implies the reduction of θ, but since the pitch angle, φ is fixed, then the angle of incidence α must decrease and C_z will drop proportionally. If the speed of rotation continues to grow over a certain value, the incidence angle can become negative. With negative incidence the value of C_z becomes negative too, so the section of the blade becomes an aerodynamic brake. The calculation procedure is based on trial and error method, until finding an acceptable twist of the blade that produces the desired progressive aerodynamic brake effect across the whole design range of wind speeds. Rotors designed with such criterion can reach very high speeds and dynamic balancing in the workshop is mandatory. The turbine should be equipped anyway with some sort of electronic control capable of reducing the electric load at the generator's terminals, in order to allow the rotation speed to quickly grow to an equilibrium point, without increasing too much the axial load on the blades. For instance, the commercial turbine of the Dutch brand *DonQi®* has been designed to reach 2.500 R.P.M., and each unit is balanced and tested in the factory for such speed. The rotor of the said turbine is made in molded plastic and its diameter is 2 m. The power output curve shows decreasing steps as the wind speed grows, with the scope of ensuring its structural integrity. Such power steps result from an electronic control system of the generator's load (Figures 4.4 and 4.5).

4.3.1.2 Fixed Pitch Rotor with Passive Stall and Constant Speed

This type of rotors are designed in such manner that the blade's sections usually work with incidence angles close to stall, and their rotational speed is kept constant by means of an external device (e.g., by connecting the

FIGURE 4.4
Photo of the *DonQi®* turbine, from the catalogue of its manufacturer.

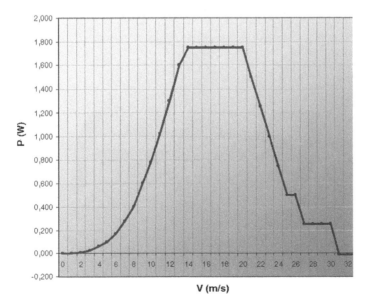

FIGURE 4.5
Power curve of the *DonQi®* turbine, from the catalogue of its manufacturer.

generator's terminal directly to the grid; by means of mechanical or aerodynamic brakes). If the wind speed grows over the nominal value, but the rotation speed is kept nearly constant, λ decreases proportionally. Decreasing λ leads the angle θ to increase, but since the pitch angle φ is fixed, the incidence angle α grows, and hence the blade stalls, reducing dramatically C_z. Such technique is called *passive stall*. The most common strategies of passive stall are the following: installing an aerodynamic brake system, driven either by centrifugal and/or aerodynamic forces (Figure 4.6); installing a purely centrifugal aerodynamic brake, like the vintage *Wincharger*® system (Figure 4.7), a disk or drum brake controlled by an electronic or centrifugal system; or a set of external electrical resistances connected in series with the generator, in order to limit its current and to dissipate the excess of energy.

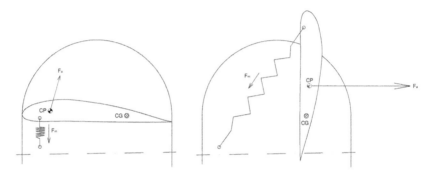

FIGURE 4.6
An example of aerodynamic brake placed at the blade's tip.

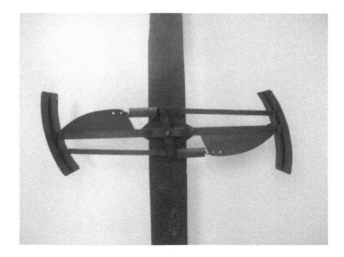

FIGURE 4.7
Detail of the aerodynamic brake system of the vintage *Wincharger*® model. Photo by the Author, with thanks to the *Museo della Civiltà Contadina* (Museum of the Peasant Culture) of Gioia del Colle (Bari Province, Italy)

4.3.1.3 Fixed Pitch Rotor Controlled by Active Stall and Variable Speed

The blades of this kind of rotors are designed to work normally with incidence angles close to the optimum, in order to obtain the maximum aerodynamic efficiency at wind speeds smaller or equal to the nominal. When the wind speed grows, the speed of the rotor is controlled by varying the resistant torque of the generator. The torque is controlled by varying the excitation current if the alternator is of the synchronous type, or by switching on external resistances, or by switching auxiliary windings that operate as electromagnetic brakes. Bringing the rotation speed below its nominal value, when the wind grows, λ falls dramatically. The decrease of λ implies the increase of the angle θ, but since the pitch angle φ is fixed, the incidence angle α grows, and the blade stalls, reducing C_z. This technique is called *active stall*, because it requires acting on the rotation speed in order to produce the stall only when necessary. Turbines controlled by active stall produce more energy along the year than those controlled by passive stall, because they work at maximum C_p with winds weaker than the nominal, which is the most frequent operation condition.

4.3.1.4 Fixed Pitch Rotor Controlled by Aerodynamic Brakes

This is one of the oldest and simplest techniques of rotor design, preferred for remote areas where technicians skilled in electronics are not available. The most famous systems are shown in Figures 4.6–4.8.

FIGURE 4.8
Wincharger® model, featuring a system of centrifugally operated aerodynamic brake and a mechanic drum brake for manually blocking the rotor. Photo by the Author, with thanks to the *Museo della Civiltà Contadina* (Museum of the Peasant Culture) of Gioia del Colle (Bari Province, Italy).

A third solution, suitable for rotors with three or more blades, consists of installing each of the blades with a different pitch angle. The individual pitch angles are optimized for a given range of wind speeds, but with the general criterion that the aerodynamic performances of the single blades are inversely proportional to the speed of rotation. The advantages of such solution are a "flatter" power output curve along a wide range of wind speeds, and a simple and reliable construction (no moving parts or auxiliary systems). The drawbacks are a lower efficiency at wind speed below the nominal, and more vibrations transmitted to the support structure. The vibrations are caused by the fact that, since each blade has a different pitch, the net axial forces will be different, hence the resultant will not fall exactly at the center of the shaft but will move along a circular path around it. This in turn will induce precession forces that will cause the rotor to yaw around a central position aligned with the wind's direction. Such a problem can be minimized by employing pair numbers of blades and installing diametrically opposite of those having the same pitch angle.

In the case of the aerodynamic brake shown in Figure 4.6, when the wind and rotation speeds are below the nominal, the aerodynamic force F_a produced by the wind tip winglet, is applied at the center of pressure CP. The said force will produce a moment with respect to its hinge, CG. The spring is pre-tensioned with a force, F_m, that produces a moment with respect to CG bigger than the aerodynamic moment. The airfoil's chord is oriented tangentially, in order to move with angle of incidence null along its circular path, producing minimum drag.

When the rotor reaches the limited rotation speed defined by the designer, the tangential speed of the blade generates an aerodynamic force on the winglet, that produces a moment bigger than the spring's one. The incidence angle increases, increasing further the moment and the drag, which tends to lower the rotation speed. If the wind speed continues to grow, the winglet will eventually flip to 90° to the tangential speed, producing a braking force equal to that of a flat plate of area A, completely exposed to the wind, whose value can be calculated with the following formula:

$$F_x = \frac{\rho}{2} \cdot 1.28 \cdot A \cdot V^2$$

F_x produces a moment much bigger than the one caused by the spring, remaining in such position until the rotation speed drops under a given value; in general, lower than the one that triggers the mechanism. Such system has the advantage of its big simplicity, both for its calculation and construction, but it is not able to completely stop the rotor, since it will resume automatically to its original position under a certain speed of rotation. When designing the support structure, it is then necessary to consider the axial thrust on the rotor turning at its minimum speed, but in high winds.

4.3.2 Variable Pitch Rotor

This kind of rotor is the standard for big turbines. In the past (until the early 1990s), it was frequent also among turbines rated a few kW, and its scope was to control the blades' pitch in order to keep the speed constant. This made possible to connect synchronous alternators in parallel to the grid, or to build stand-alone systems without the risk of overcharging the batteries and auxiliary electric components. Including a system of pitch variation of the blades in the design of a wind turbine adds more mechanical complexity to the rotor but, on the other hand, provides a more stable operation, both of the rotation speed and power output, and protects the turbine in case of excessive wind, by placing the blades in "flag position." Three methods exist allowing to control the pitch of the blades, as will be explained in the following sections.

4.3.2.1 Variation of the Pitch by Means of Servomechanisms

The usual servomechanism employed to vary the blades' pitch in big turbines is a hydraulic piston mounted concentrically with the shaft, or a linear electric actuator. Such solution is not applicable to small turbines because of its big mechanical complexity and the need to include a microprocessor-controlled system to govern the hydraulic circuit.

Control algorithms must be programmed on purpose for the said scope based on *fuzzy logic*, which allows adapting the operation to produce the maximum power under any wind condition, considering at the same time the structural resistance of the turbine and its support structure, and the electric load on the generator.

4.3.2.2 Variation of the Pitch by Means of Centrifugal Force

The general principle of operation is shown in Figure 4.9 and involves placing some centrifugal masses concentrically to the shaft. An increase in the wind speed will increase the rotation speed, which in turn increases the centrifugal force on the masses. Such forces will displace the masses from their initial position of equilibrium, and such displacement is then transmitted to the blades' joints, causing them to vary their pitch. The pitch is varied in order to reduce the incidence angle and consequently C_z and C_x, reducing then the power output and, at the same time, the thrust on the hub.

Figure 4.9 illustrates the details of the simplest centrifugal system in which the masses are independent, one for each blade. If the blade's hinge axis coincides with the CP of the airfoil, then the system is in equilibrium when the elastic moment produced by the spiral spring is equal to the moment produced by the centrifugal force on the counterweight of mass m:

$$M = k \cdot \phi = m \cdot \omega^2 \cdot r \cdot a$$

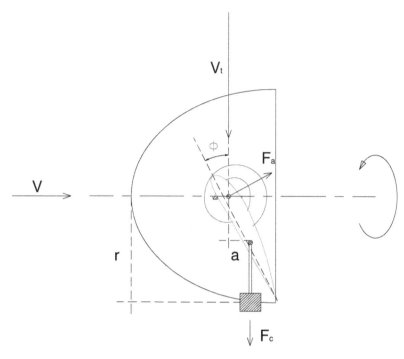

FIGURE 4.9
Sketch of pitch control by independent centrifugal masses attached to each blade.

where:
  M = Elastic moment
  k = Elastic constant of the spring (N/°)
  φ = Angular deformation of the spring, coincident with the pitch angle (mass directly coupled to the blade) or directly proportional to it (mass coupled through gears) (°)
  m = Centrifugal mass (kg)
  r = Distance from the center of gravity of the centrifugal mass to the center of the rotor's shaft (m)
  ω = Angular speed (rad/s)
  a = Lever arm of the centrifugal force to the blade's hinge (m)

By definition:

$$\Omega = \frac{\lambda \cdot V}{R}$$

So we get:

$$\phi = m \cdot \frac{\lambda^2 \cdot V^2}{k \cdot R} \cdot r \cdot a$$

Figure 4.10 shows one of the many possible systems to control the pitch of the blades by means of centrifugal masses connected via a crown and pinion, known as *Jacobs system*.

As an alternative, the centrifugal masses can act on single elements of blade (for instance, by turning the blade's tip as an aerodynamic brake), or on *spoilers* or *flaps*, in order to induce the stall, or by varying the geometry of the airfoil (Figure 4.11).

The disadvantage of centrifugal systems is that the masses turn in a vertical plane. Consequently, they are subject to the maximum acceleration at the lowest point of their trajectory, because the centrifugal force adds to the gravity and, on the contrary, the force is minimum at the highest point of their trajectory, where the gravity and centrifugal forces have opposite senses. In the intermediate positions the forces vary according to a sinusoidal law. Consequently, the pitch angle of the blade will undergo small oscillations around an equilibrium value. Such oscillations induce in turn unavoidable

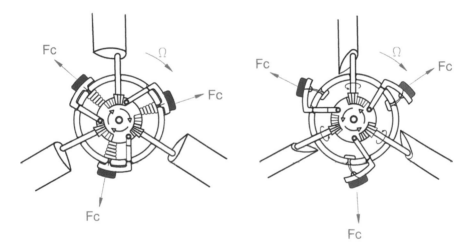

FIGURE 4.10
Operation principle of the centrifugal control of *Jacobs* turbines. Adaptation of drawings from Le *Gourière*, with thanks to prof. Arch. Mario Barbaro for the improvements.

FIGURE 4.11
A centrifugal mass, placed inside the blade, displaces a wedge that in turn moves the spoiler.

vibrations and fatigue stresses. The *Jacobs system* and similar ones brilliantly solve the problem, because the centrifugal masses and the blades are all connected with a gear crown, so the pitch of the blades results from the average of the centrifugal and gravitational forces on the three masses. The reliability of the Jacobs system is demonstrated by the fact that one of the units built by the company (closed when the owner retired in the 1980s) worked uninterruptedly from the 1930s to the end of the twentieth century in an American base in Antarctica. According to the inventor's sons, it still worked when the base was dismantled. On the other hand, the *Jacobs* system is difficult to build with off-the-shelf components.

Another limitation of centrifugal control systems is their inability to stop the rotor completely. For this reason they are generally combined with other systems of protection, for instance, yaw controls that flip the rotor's plane parallel to the direction of the wind when this latter is too strong.

4.3.2.3 Pitch Control by Aerodynamic Moment

Figure 4.12 shows the principle of functioning of such system. In general, pitch controls by aerodynamic moment are rare, because they are not able to control the rotation speed with precision. On the other hand, they protect

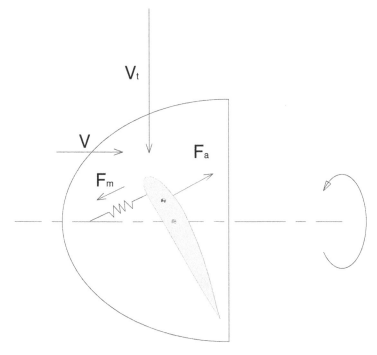

FIGURE 4.12
Pitch control by aerodynamic moment.

very well turbines against high wind speeds, because they are able to automatically "flag" the blades. The ideal solution is equipping the rotor with a compound system, centrifugal and aerodynamic, in order to obtain protection against both high winds and high rotation speeds.

4.3.3 Yaw Systems and Variation of the Exposed Surface

Systems of this kind are very easy to build, but their function is protecting the turbine against strong winds rather than controlling its rotation speed or its power output.

Mechanical switch-off systems usually have a hinged tail or an auxiliary tail placed at 90°. The latter system is considered obsolete, and still can be found in some American multi-blade turbines for pumping water. Figures 4.13 and 4.14 illustrate the operation principle of one of the possible solutions. The turbine is hinged to its attachment to the mast. Under normal wind conditions, the capsizing moment with respect to the hinge O, caused by both the force on the hub, F_m and the aerodynamic force on the horizontal rudder, F_a, is smaller than the stabilizing moment generated by the weight of the turbine, the eventual gears and the generator, G.

By dimensioning the horizontal rudder on purpose, it is possible to generate an aerodynamic force causing a capsizing moment bigger than the stabilizing moment, but only when the wind speed will be higher than a value

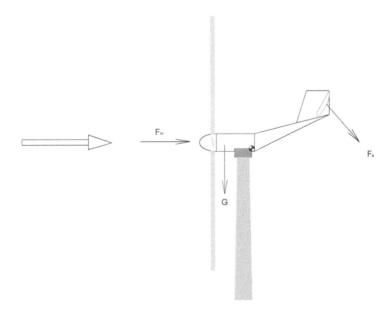

FIGURE 4.13
Example of a mechanical protection device able to yaw the rotor off the wind's direction, when running under normal operational conditions.

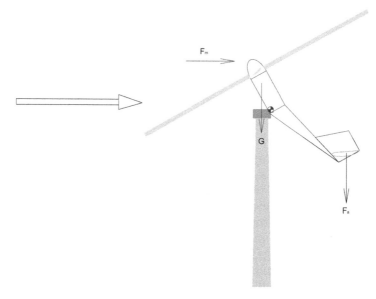

FIGURE 4.14
The same device under strong wind conditions.

predefined by the turbine's designer. Under such conditions the turbine will rotate around the hinge *O*, presenting a minimum exposed area and avoiding eventual damages to the rotor or the generator.

Turbines with down-wind rotor offer the possibility to hinge the blades in such a way that the axial force will tend to collapse them, reducing the exposed area (Figure 4.15). The collapsing moment is counterbalanced by a simple spring. The increase of the capsizing angle, β, reduces the rotors' radius. The exposed area varies according to a sinusoidal law:

$$A = \pi \cdot R^2 \cdot \cos^2 \beta$$

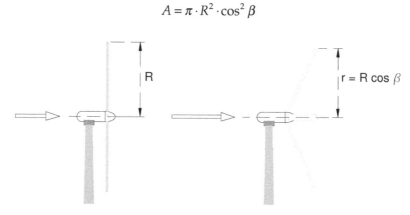

FIGURE 4.15
Control system of the exposed area for down-wind rotors.

4.4 Selecting the Materials and Techniques for the Blades' Manufacture

4.4.1 Wood

This is one of the most sustainable materials, easy to work with, widely available and having good mechanical properties. Nevertheless, it has a series of drawbacks, as for instance, high weight for a given stiffness and resistance, sensitive to moisture, low superficial hardness (abrasion if the turbine is installed in dusty areas), requires yearly maintenance. Wood is easy to work with using hand tools and does not require specialized personnel, so it is the material of choice when the scope is building a single turbine, or small series, in rural areas with scarce resources.

Plywood is preferable to massive wood, but it must be protected with an adequate coating anyway. Flat-convex airfoils, like the Clark Y, or the Drela AG03, facilitate the construction of flat blades (variable chord, no twist) by gluing together multiple layers.

4.4.2 Fiber Reinforced Plastic Resin (FRPR)

This is the material of choice for manufacturing the blades of commercial wind turbines. It requires an investment in molds, but allows serial production at lower unit costs, and provides optimum mechanical resistance. The high elasticity of Fiber Reinforced Plastic Resin (FRPR) allows the design and construction of small self-regulated rotors. The blades are designed in such a way that they will deform in high winds, changing the rotor's geometry, for instance, its twist. High-performance turbines employ plastics reinforced with carbon fiber, because of the excellent mechanical properties of such composite material. Nevertheless, carbon fiber is more expensive than glass fiber, and the increase in mechanical resistance that it can provide does not justify its higher cost in the case of small turbines.

4.4.3 Aluminum Alloys

Aluminum allows the production of extruded profiles and sand-cast or plaster-cast pieces. Extruded profiles allow high production volumes, but require special equipment and tooling. There are some specialized companies that supply such profiles, for instance, http://www.thebackshed.com/windmill/Trade/AlBladeOrders.asp and https://se110802384.fm.alibaba.com/?spm=a2700.8304367.shopsigns.2.69ef10c3ZTaAuY. The drawback of extruded profiles is that shaping the turbine's blades requires manual labor, applying torsion to obtain the required twist at each section of the blade, and the resulting blades are not exactly equal. Alternatively, the rotor can be produced with constant-chord, flat blades at the expense of sacrificing aerodynamic performance. Another alternative is shaping the blade from aluminum sheet, either on a wooden form or employing prefabricated systems, as for instance, the solution proposed by *Windknife®* http://www.windknife.com/. The joint between the blade and the turbine's shaft must be done carefully, avoiding direct contact between the aluminum blade and steel bolts or other parts. Always use plastic washers and inserts, in order to prevent corrosion points caused by the contact of aluminum with other metals.

4.4.4 Sail Rotor

The drawback of sails is their low C_z/C_x ratio (nearly 20 for infinite wingspan), and the consequently low aerodynamic efficiency of the rotor built with them. It is possible to partly offset such problem by building a bigger rotor, since the difference of cost would be insignificant. Sails require some form of pitch control, in order to keep an incidence angle that ensures they will maintain their shape and will not flutter. Fluttering not only degrades the sail's aerodynamic features, but will damage it if it persists for a long time. Ancient windmills in the Mediterranean area used to employ simple triangular sails attached to radial wooden poles and ropes (Figure 4.16).

Some more modern experiences have re-proposed sail rotors, a.k.a. *Cretan windmills*, for slow wind turbines pumping water, as the one depicted in Figure 4.17.

FIGURE 4.16
Windmill of Pasico in operation during the XIV Festival of the Windmills in Torre Pacheco, Campo de Cartagena, Spain, 27th April 2014 (http://www.torrepacheco.es/turismo/rutas-turisticas/los-molinos-de-viento). With thanks to Mr. José Antonio Rodríguez for the authorization to reproduce this photo.

FIGURE 4.17
A modern version of the Cretan windmill pumping water in Greece, 19th May 2013. With thanks to Mr. Neil Harvey for the authorization to reproduce his photo.

The growing popularity of windsurf has led to the development of fully battened, semi rigid profiled sails. Profiling the mast adequately, and battening the sail (Figure 4.18), results in a flexible blade with better aerodynamic performances than the conventional sails employed in traditional windmills. The reason why such kind of sails is not very diffused in the nautical sports market is simply that they are considered "anti-sportive," but from the point of view of the small wind turbine builder they are a cheap and valid solution for the construction of small rotors. The usual materials employed in commercial sail making are *Dacron*® or similar polyester fabric, or laminated *Mylar*® (semi-rigid sails). Canvas or tarpaulin can be a suitable solution if the

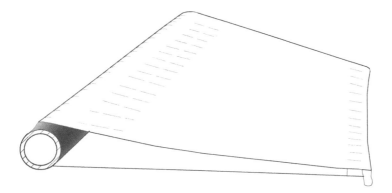

FIGURE 4.18
Double-layer sail wrapped around a tubular mast. It is highly advisable to include several battens, allowing the sail to keep the shape as close as possible to the base airfoil employed for the calculations. With thanks to prof. arch. Mario Barbaro for digitizing the Author's handmade sketch.

turbine must be built and operated in low-income rural areas. Polyethylene, PVC or polypropylene foils are inexpensive and widely available commodities, allowing the cheap construction of semi-rigid sails, although their lifetime may be shorter than a couple of years. The following sequence allows the analysis of a sail rotor's aerodynamic features:

1. Choose a concave–convex airfoil, whose geometry can be approximated as close as possible when designing the mast and the sail combination. Some examples of such profiles are the GOE 417, or the GOE 417A, o the Selig S1091A.
2. Design the blade with the aerodynamic coefficients of such airfoil, imposing arbitrarily linear or null twist, and linear variation of the chord (triangular or trapezoidal sails), or even constant chord (rectangular sail).
3. Check the rotor's resulting C_p for different combinations of wind and rotational speed.
4. Plot the curve of C_p as a function of λ (or of P as a function of V) and decide if it is acceptable, otherwise repeat the procedure from the beginning with another airfoil, or with different twist and chord laws.

4.4.5 Wooden Frame Covered with Tarpaulin, or Plastic Foil, or Thin Metal Sheet

This is a hybrid solution between a conventional blade and a sail. At the dawn of aeronautics, airplane wings used to be built with this technique. It is an interesting solution for the manufacture or single turbines, using simple hand tools and cheap materials.

4.5 Practical Examples

4.5.1 Low-Cost Rotor with Profiled Sail

4.5.1.1 Definition of the Problem

The mission is to design a low-cost wind turbine for basic electricity supply to rural inhabitants in a developing country. An old alternator, $12\,V \times 60\,A$, recovered from a dismantled car, is available. The minimum rotation speed of the alternator is estimated to be 1,200 R.P.M. From a preliminary anemometric study on site, the speed of the wind that maximizes the total annual quantity of energy is 5.5 m/s; hence, the turbine must reach its nominal power at the said wind speed.

4.5.1.2 Pre-dimensioning

Apply the formula presented in Section 4.2.1.

$$D = \sqrt{\frac{P}{0.20 \cdot V_1^3}}$$

$$D = \sqrt{\frac{12 \cdot 60}{0.20 \cdot 5.5^3}} = 4.92 \text{ m} \approx 5\text{m}$$

Assume the number of blades $z = 2$, for the sake of manufacturing simplicity and least quantity of material and labor. Since it must be a fast turbine, but the construction with locally available materials and rudimentary techniques does not allow to reach a high fineness ratio of the airfoil (rugosity, offset from the theoretical geometry, random deformations, etc.), assume as a conservative value $\lambda = 5$. Consequently:

$$N = \frac{\lambda \cdot 60 \cdot V}{\pi \cdot D}$$

$$N = \frac{5 \cdot 60 \cdot 5.5}{\pi \cdot 5} = 105 \text{ R.P.M.}$$

It is necessary to foresee a multiplication gear between the turbine's and the alternator's shaft. The multiplication ratio is then:

$$Z = \frac{N_2}{N_1} = \frac{1200}{105} = 11.4$$

Such multiplication can be made by means of a pulley and belt system, friction wheel, pinion and chain, or gearbox. Since the required power is small (about 900 W at the shaft of the turbine) the most suitable solutions for our case are either a system of pulleys and belt (recovered from dismantled vehicles) or a pair of friction wheels built with plywood and coated in rubber, or a chain and pinion recovered from an old bicycle.

4.5.1.3 Yaw System

We will adopt the classical vane described in Section 4.2.2.1, Figure 4.1.
Its calculation does not require explanations. Define arbitrarily $e = 0.5$ m and $L = 2$ m (eventually, the choice must be adapted as a function of the alternator's dimensions and of the available multiplication system)

$$s = 0.16 \cdot \frac{\pi \cdot D^2}{4} \cdot \frac{E}{L}$$

$$s = 0.16 \cdot \frac{\pi \cdot 5^2}{4} \cdot \frac{0.5}{2} = 0.78\,\text{m}^2$$

The shape is irrelevant. For constructive simplicity and minimum use of material, and in order to provide a minimum of aesthetics the finished turbine, it is possible to make the vane's tail with the shape of a triangle. Take a metal plate of $1 \times 0.78\,\text{m}$ and cut it along the diagonal. Weld the two resulting triangles to the tail stub, forming a bigger triangle, $1\,\text{m}$ high by $1.56\,\text{m}$ base. The distance between the barycenter of the triangle and the joint of the turbine to the support structure must be L.

4.5.1.4 Selection of the Airfoil

In this example, we postulate the construction of the rotor employing rigid profiled sails. The easiest way to build the core of the sail is to employ a cambered metal sheet. Low-speed turbines largely employ such solution, but in our example we need to build a fast turbine, hence a simple cambered plate is not suitable because its f is too low. We will then adopt an E377 MOD airfoil, whose data are included in Chapter 12. The construction of a 2.5 m long blade with the said airfoil is relatively easy. First of all, it is necessary to build a wooden template with the camber line of the airfoil, and then shape a metal sheet on it. Finally, it is necessary to build the airfoil's nose that will be applied to the leading edge. The nose can be built with another curved plate, or with wood or plastic, or even with polyurethane foam (this will require building a separate mold).

4.5.1.5 Dimensioning of the Blade

Considering the geometry of the chosen airfoil and in order to keep the construction as simple as possible, it is natural to design the blades of the turbine with constant chord and twist angle. Under such conditions, the blade's performance will result sub-optimal, but such problem is compensated by a bigger diameter of the rotor resulting from generously rounding up (Section 4.5.2). Suppose arbitrarily that the incidence angle in the center of the blade is such, that the fineness coefficient, f, is maximum. From the aerodynamic data of the airfoil E377 MOD presented in Chapter 12, the resulting α and its corresponding C_z are:

$$\alpha = 8° \rightarrow C_z \approx 1.5$$

Suppose then that the average value of C_z between the blade's joint to the shaft and its tip is 1.5.

According to Equation 3.21, the optimum angle θ is:

$$\theta = \frac{2}{3}\text{arctg}\left(\frac{1}{\lambda_r}\right)$$

At the central point of the blade, the value of λ_r is half of the value of λ assumed during the pre-dimensioning stage, hence, in our example:

$$\theta = \frac{2}{3}\arctg\left(\frac{1}{2.5}\right) = 14.5°$$

According to Equation 3.2, we get:

$$\sigma_l \cdot C_z = 4 \cdot (1 - \cos\theta) = 4 \cdot (1 - \cos 14.5°) = 0.128$$

Replacing the value of C_z chosen before, we obtain:

$$\sigma_l = \frac{0.128}{1.5} = 0.085$$

From the definition of σ_l, we can finally calculate the value of l:

$$\sigma_l = \frac{z \cdot l}{2 \cdot \pi \cdot r} \quad \rightarrow l = 2\pi \cdot 1.25 \, \text{m} \cdot \frac{0.085}{2} = 0.335 \, \text{m}$$

Since it is impossible to build a blade starting from the geometric center of the shaft, in general its length is arbitrarily defined as 90% of R. In our example, the resulting rectangular blade will have

$$L = 0.9 \, R = 0.9 \cdot 2.5 \, \text{m} = 2.25 \, \text{m}$$

The chord of the airfoil will be constant, $l = 0.335$ m, and the blade must be attached to the shaft in such a way that its chord will form an angle φ with respect to the rotation plane, resulting from the difference shown in Figure 3.5:

$$\varphi = \theta - \alpha = 14.5° - 8° = 6.5°$$

4.5.2 Design of an Optimum 3-Bladed Rotor for Electrical Generation

4.5.2.1 Definition of the Problem

We desire to design an optimum rotor for electrical generation in a site where winds are weak. The required power at the shaft is 1,500 W, and the nominal wind speed for producing such power is 5 m/s. The reader will find the file *windmill-design-2017.xlsx* in the download section. It allows the amateur to easily design, analyze the performance and check different construction alternatives (optimum blade, optimum chord with no twist, both linear chord and twist, constant cord with no twist, and in all cases varying the blade's pitch angle, etc.). The spreadsheet provides a good accuracy of results but the reader must keep in mind some limitations:

- The profile chosen is NACA 0012. This choice was dictated by the fact that the Author has aerodynamic data for this profile beyond the stall point. This is a very important feature for the analysis of the windmill's performance, since, unlike an aircraft propeller, a windmill must operate in a wide range of speeds. It's difficult to find such data for other profiles. Furthermore, NACA 0012 is a symmetrical and easy to build profile. With negative angle of attack the C_z value is the same of the positive case, but with opposite sign. The C_z and C_x data of the profile were tabulated with a step of 0.5° through linear interpolation in the "usual" range of angles of attack. In the stall range and over, the precision is lower (data derived graphically and furthermore linearly interpolated!). The reader can simply replace the data with those of any other profile, putting attention in not changing the position of the cells, adding or cancelling cells.

- In some cases, it's possible that the table displays C_P bigger than the theoretical, what means that the rotation speed assumed for the calculation is impossible. A warning message will display in such cases.

- There's another source of error which will be evident while eventually testing a model designed with this tool: C_z and C_x vary with the Reynolds number, especially when operating in the stall range or with very small turbines in weak winds. The tabulated data are valid for Re = 1.8×10^6. This means that for weak winds and/or low-speed rotations the real output will be much lower than the calculated one. In the upper speed range, the influence of Re is not so big.

- The λ factor is valid in the range $0.01 < \lambda < 13$. Out of this range, the results will contain significant errors.

- It is assumed that the first quarter of the blade's length will have no profile, although the aerodynamic drag of this portion (usually a tube) is not considered in the calculations.

- The remaining 75% of the blade's length is divided into 30 points. The spreadsheet performs a numerical integration with a small step (2.5% of the total rotor radius). Please, note that for a windmill having 1 kW output, this means that each "differential" section is about 4 cm long, and the author considers that the precision should be enough up to 10 kW power. If more precision is needed, the step's value should be reduced accordingly, i.e., the length will be divided into 50 or more discrete intervals, which means adding more rows with the "copy down" function.

- Negative torque and power means that the incidence angle on the profile is negative. In this condition the blade is acting as a brake instead of producing power.

4.5.2.2 First Step: General Size of the Turbine and Orientation Vane, Generating the Optimum Blade Shape

The first sheet of the file calculates the diameter and consequently the dimensions of the orientation vane, as well as generates the optimum chord and local pitch angle for each section of the blade. It also plots an approximate (not in scale) planform and twist view of the blade, as well as the mechanical loads on each section. The only inputs necessary are: the desired power, the wind speed, the density of air (in case the turbine is meant for operating at high altitude or in extreme temperatures, otherwise leave the default value), the number of blades, the desired angle of attack at each section (usually the angle that provides maximum f), and the desired λ. Hint: try different combinations of angle of attack, number of blades and λ, and check how C_P varies with them.

4.5.2.3 Second Step: Performance of the Optimum Blade

This sheet takes as fixed values the ones calculated in the first step, and allows to calculate P, λ, and C_P using as variables V, the pitch angle θ, and the rotation speed (in R.P.M.). It is then possible to analyze different operational strategies: active or passive stall, pitch control, and obtain the C_P versus λ and P versus V curves.

4.5.2.4 Third Step: Evaluating Alternatives

The successive sheets allow to compare the optimum blade to different scenarios: constant cord and pitch angle (blade made with a single plank of wood); linear variation of the chord and local pitch (trapezoidal-profiled sail); optimum chord variation with no twist (blade made of plywood), linear variation of the chord with optimum twist (aluminum or plastic sheet on a ribbed structure). The file contains some examples and explanations.

Bibliography

Abbot I., Von Doehnhoff A., *Theory of Wing Sections*, Dover Publications Inc., New York, 1958.

Le Gourière D., *L'Énergie Éolienne—Théorie, Conception et Calcul Practique des Installations*, deuxième édition, Eyrolles, Paris, 1982.

Manwell J.F., Mcgowan J.G., Rogers A.L., *Wind Energy Explained: Theory, Design and Application,* 2nd edition, John Wiley & Sons Inc., Chichester, UK, 2010.

Rosato M., *Diseño de Máquinas Eólicas de Pequeña Potencia,* Editorial Progensa, Sevilla, Spain, 1992.

Tangler J.L., Somers D.M., *NREL Airfoil Families for HAWTs,* National Renewable Energies Laboratory, Colorado, 1995.

Selig M., *PROPID, A Free Software for the Design of Horizontal Axis Wind Turbines,* University of Illinois Applied Aerodynamics Group, 2012, downloadable from http://m-selig.ae.illinois.edu/propid.html.

5

Practical Design of Aerodynamic
Action Vertical Axis Wind Turbines

5.1 General Considerations about Vertical Axis Turbines

Vertical axis wind turbines are classified into two families: aerodynamic action turbines (*Darrieus* type, the subject of this chapter), reaction turbines (*Savonius* type, see Chapter 6), and mixed action (*Lafond* type, not treated in this book). *Darrieus* turbines have a vertical axis and their operation is based on the lift principle; their name derives from the inventor, *George Darrieus*, who patented them in 1931. Such class of turbines is capable of reaching values of C_P comparable to those of horizontal axis turbines, but slightly smaller. The name *Darrieus* applies to two classes of vertical axis turbines: the original type—patented by Darrieus—informally called *"eggbeater"* because of its characteristic shape, and the "H-type"—featuring straight blades and cylindrical elevation—which is easier to design and build than the first model.

A more modern variant is called *Gorlov turbine*. This one is nothing but an H-type *Darrieus* turbine in which the blades are helicoid shaped instead of being straight. All the variants of the *Darrieus* family run because of the lift created by the profile of the blades. Figure 5.1 shows a model of the original *Darrieus* typology, in 2-blade version. In practice, big vertical axis turbines have never encountered the favor of investors because of their lower efficiency and higher operational complexity for a given cost. On the contrary, they are very popular among engineers, architects, and private citizens, more because of aesthetical and marketing hypes than for objective technical reasons. The most common opinion, very arguable because not completely founded, is that vertical axis wind turbines are easier to integrate in urban environment than horizontal axis wind turbines.

Darrieus turbines have two big limitations that have hampered their massive diffusion, especially in the market of small installations.

The first limitation is that such turbines operate in a narrow interval of λ values, ranging from four to eight, and furthermore they are not capable of

FIGURE 5.1
Two-blade *Darrieus* turbine installed in Aguinaga, Gran Canaria, Spain. Photo by the Author.

starting autonomously. The second limitation is that their C_P is intrinsically lower than that of horizontal axis turbines, because of their particular operation principle. Their constructive simplicity is neutralized by the need of installing a start motor or another start system, in order to bring the rotor to the necessary speed to keep it in stable rotation. Several solutions exist that can overcome the said problem. One of them consists of installing a reaction turbine (usually a *Savonius* turbine) inside the *Darrieus* turbine (Figure 5.2). The *Savonius* turbine should have such dimensions that its motor torque is able to bring quickly the blades of the *Darrieus* rotor to their nominal speed, so that they can generate enough lift to keep a stable rotation. Once said stability condition is reached, the ancillary turbine is mechanically uncoupled from the main shaft, by means of a clutch or any similar mechanism. As an alternative, since the *Savonius* turbine works with $\lambda < 1$, it is possible to operate a *Savonius* turbine rigidly coupled with the main shaft of the *Darrieus* turbine, on the condition that the radius of the *Darrieus* turbine is five or six times bigger that the radius of the *Savonius* one. In the hybrid *Darrieus–Savonius* configuration, the *Darrieus* rotor tends to accelerate in strong winds, while the *Savonius* rotor has a tendency to limit the speed, like an aerodynamic break, providing then a more regular operation of the system. Another start system relies on the generator itself, which serves as electric motor during the start phase until the rotation speed becomes stable. Though such a method is easy to implement, it is not applicable everywhere, because it requires a connection to an external electric grid and hence, it precludes the stand-alone operation of the turbine.

FIGURE 5.2
Darrieus turbines with automatic start by means of auxiliary *Savonius* turbine, installed in Taiwan. Free license photo by Mr. *Fred Hsu*.

5.2 Simplified Theory of the Darrieus Turbines

A theory of *Darrieus* turbines based on *Glauert's* theory of horizontal axis turbines ignores the effect of the trail vortexes of the upstream blades on the downwind ones. Hence, such theory tends to yield too optimistic values of C_P. Sophisticated numeric models are available for the calculation of the *Darrieus* rotor, but these are out of the possibilities of the average small-scale constructor. For such reason our treatise will be limited to the classical method, developed by the Canadian researcher *R.J. Templin* in the 1970s, and later modified by the French *Le Gourière*. Said method provides acceptable results, validated in the wind tunnel and comparable to the tests carried out in real scale by the *Sandia Laboratories* in the 1980s.

The driving forces of the *Darrieus* turbines can be described in detail by means of the "element of blade theory" and with the help of Figure 5.3. There are two components of the speed to be considered: the tangential speed of

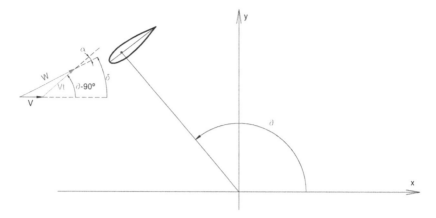

FIGURE 5.3
Components of the apparent wind speed acting on a differential element of a *Darrieus* rotor blade in a generic point of its trajectory.

the airfoil around the shaft, $V_t = \Omega \cdot R$ (in light grey), which is always parallel to the airfoil's chord, and the speed of the wind across the rotor, V (in black). The latter is assumed constant in direction, module, and sense. The first problem to solve is to define what the value of V is. The second problem is to define a correction coefficient in order to account for the aerodynamic interference caused by the upwind blades on the downwind blades. Observing Figure 5.3, we can easily realize that the supposition of constant wind direction downwind of the blade is not completely true, because the direction of the airflow is influenced by the aerodynamic action of the airfoil. We can assume that our suppositions are true enough in the blade path from $\theta = 0°$ to $\theta = 180°$ in Figure 5.3, but too optimistic in the rest of the path.

The resultant of the composition of the speed vectors V_t, tangential, and V, across the rotor, is the speed of air relative to the airfoil, also known as apparent speed of the wind, W, which is a function of the position of the blade along its circular trajectory.

The angle of attack, α, determined by the resultant W and by the airfoil's chord, will then vary with the position of the blade along its circular path around the shaft. The torque generated by a *Darrieus*-type rotor is hence pulsating, and in order to make it uniform it would be necessary to build the rotor with many blades. But increasing the number of blades, the bigger quantity of vortexes caused by the blades in the upwind half of the trajectory will hamper the performance of the other blades in the downwind half, and this is the main inconvenience of such type of rotor. In practice, rotors with three or four blades will give the best results.

Observing again Figure 5.3, we can intuitively realize that, in both the positions $\theta = 90°$ and $\theta = 270°$, $\alpha = 0°$. In said positions, the lift is hence null and does not produce useful torque. The drag force, in turn, always generates a negative component of torque, which always tends to slow down the blades. In practice, symmetric profiles are always necessary for the manufacture of *Darrieus*-type turbines, because negative angles of attack produce negative lift, but this latter has the same module of its positive homologue.

In order to produce useful work, the torque produced by the lift forces, must be greater than the torque generated by the drag forces. Such condition is valid only in a narrow range of values of λ. Out of such range, the torque will be too small, or even negative. Said concept is easier to understand if we analyze the incidence angle and the intensity of the apparent wind from the point of view of an imaginary observer moving with the airfoil. If we place our reference axes in such a way that the x-axis coincides with the chord of the airfoil, the component V_t becomes a constant in x, and the component V can be represented as a rotational vector with its origin at the end of V_t. In physical terms, this means that, instead of analyzing the relative motion of the blade around the shaft, we imagine the blade as being fix, and the wind turning around of it with angular speed Ω. Since the motion is always relative, there will be no difference between the results obtained in one way or in the other, but the graphical representation and the explanations are much simplified assuming the chord of the airfoil as reference axis.

Figure 5.4 shows the situation as seen from the airfoil's point of view. If $V_t \gg V$, we can deduce that then the incidence angle α is always acute. In particular, if $\alpha < 10°$, then the airfoil will never stall. This explains why the Darrieus turbines are able to turn stably as long as $\lambda < 4$: the arc tangent of 0.25 corresponds to an angle of 12° and in general, all symmetric airfoils are close to the critical point, or already stalled, with such incidence angle. In the opposite case, when $\lambda > 8$, the maximum incidence angle is 7°. All symmetric

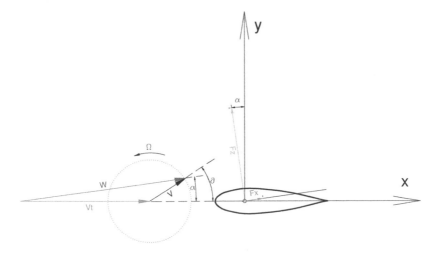

FIGURE 5.4
Speed of the apparent wind and forces relative to the chord of the airfoil, defined as coincident with the x-axis.

profiles produce little lift at small incidence angles, in other words, the blade will generate only small aerodynamic lift along most of its path, but the drag forces acting on it remain constant and hence, the total C_p of the rotor falls drastically.

Compared to horizontal axis turbines, *Darrieus* turbines have the drawback of needing a start system, but their advantage is not requiring any protection against excessive rotation speed: it is enough to detach the generator and to leave the turbine turning freely, since its speed will be self-limited. For precaution, Darrieus turbines often have an additional centrifuge or aerodynamic protection mechanism that will trigger at high speed causing some breaking effect. In other cases, the break is magnetic or mechanic.

From Figure 5.4, we can deduct all the necessary elements for the complete analysis of the operation of an "H-type" *Darrieus* rotor according to the theory of the element of blade:

$$\tan(\alpha) = \frac{V \cdot \text{sen}(\theta)}{V_t + V \cdot \cos(\theta)}$$

$$W^2 = (V \cdot \text{sen}\theta)^2 + (V_t + V \cdot \cos\theta)^2$$

$$V_t = \Omega \cdot R$$

$$F_z = \frac{\rho}{2} \cdot C_z \cdot l \cdot h \cdot W^2$$

$$F_x = \frac{\rho}{2} \cdot C_x \cdot l \cdot h \cdot W^2$$

Remember that l is the chord of the airfoil and h the length of the blade. The designer, based on the considerations described in next Section 5.3.3, will predefine such dimensions arbitrarily.

The next step is decomposing the forces F_z and F_x in their tangential and radial components, called respectively F_t and F_r. The first will generate a positive average torque for λ speeds in the range 4–10 (depending on the profile), while the second can be either centrifugal or centripetal, depending on the position of the blade along its trajectory. Since the angle of incidence is known for any point of the trajectory, the calculation of F_t and F_r is straightforward:

$$F_t = F_z \cdot \text{sen}\alpha - F_x \cdot \cos\alpha$$

$$F_r = F_z \cdot \cos\alpha + F_x \cdot \text{sen}\alpha$$

We can further simplify the last two expressions by introducing the coefficients of radial force, C_r, and tangential force, C_t, already described in Section 2.3.2.3 and repeated here for easier reading:

$$C_t = C_z \cdot \text{sen}\,\alpha - C_x \cdot \cos\alpha$$

$$C_r = C_z \cdot \cos\alpha + C_x \cdot \text{sen}\,\alpha$$

Hence, we have now two expressions that are easier to calculate:

$$F_t = \frac{\rho}{2} \cdot C_t \cdot l \cdot h \cdot W^2 = q \cdot C_t$$

$$F_r = \frac{\rho}{2} \cdot C_r \cdot l \cdot h \cdot W^2 = q \cdot C_r$$

The product of F_t for the radius R is the instantaneous driving torque, M, produced by the blade in the point being analyzed. We will call M_t the total driving torque, which is the average of the instantaneous torque calculated along the circular trajectory, multiplied by the number of blades, z.

$$M_t = \bar{M} \cdot z = \frac{z}{2\pi} \cdot \int_0^{2\pi} q \cdot C_t \cdot d\theta$$

The power is simply the product of the total torque by the angular speed Ω.

$$P = M_t \cdot \Omega$$

In order to calculate M and therefore P, we need to know the speed across the rotor, V, which is unknown *a priori*. The speed V can be expressed as a function of the speed of the wind upstream of the turbine, called V_1, following a reasoning similar to the one already explained when demonstrating Betz's Theorem.

The axial force exerted by the wind on the rotor, F_a, is the projection of the components F_t and F_r on the direction of V, hence:

$$F_a = F_r \cdot \text{sen}\theta - F_t \cdot \cos\theta = q \cdot (C_r \cdot \text{sen}\theta - C_t \cdot \cos\theta)$$

The average of F_a along a circular trajectory can be calculated as:

$$F_{a\,total} = F_a \cdot z = \frac{z}{2\pi} \cdot \int_0^{2\pi} q \cdot (C_r \cdot \text{sen}\theta - C_t \cdot \cos\theta) \cdot d\theta$$

The theory of the blade element does not allow us to calculate the value of V. We can hence base it on Betz's theory, according to which the axial force on the rotor is given by:

$$F = \rho \cdot A \cdot V \cdot (V_1 - V_2)$$

Where:
A = exposed area (in the case of the H-type *Darrieus* turbine $A = R \cdot h$)
V_1 = speed of the wind upstream of the turbine
V_2 = speed of the wind downstream of the turbine

Since, according to Betz's theory:

$$V = \frac{V_1 + V_2}{2}$$

The axial force is then:

$$F = \frac{\rho}{2} \cdot A \cdot (V_1 + V_2) \cdot (V_1 - V_2)$$

$$F = \frac{\rho}{2} \cdot A \cdot (V_1^2 - V_2^2)$$

Equalling this last expression with the one obtained from the theory of the forces on the blade element, we get the following expression:

$$\frac{\rho}{2} \cdot A \cdot (V_1^2 - V_2^2) = \frac{z}{2\pi} \cdot \int_0^{2\pi} q \cdot (C_r \cdot \text{sen}\theta - C_t \cdot \cos\theta) \cdot d\theta$$

Replacing q with its equivalent expression and simplifying, the former equation reduces to the following:

$$(V_1^2 - V_2^2) = \frac{l \cdot h \cdot z}{2\pi \cdot A} \cdot \int_0^{2\pi} W^2 \cdot (C_r \cdot \text{sen}\theta - C_t \cdot \cos\theta) \cdot d\theta \qquad (5.1)$$

From the definition of the speed drop coefficient a, we can then write:

$$V_1 = \frac{V}{1-a}$$

$$V_2 = V_1 \cdot (1 - 2a)$$

and from Betz's Theorem, the following condition must be fulfilled:

$$V = \frac{V_1 + V_2}{2}$$

With the last three equations, it is finally possible to develop the first term of Equation 5.1 as follows:

$$V_1^2 - V_2^2 = (V_1 - V_2) \cdot (V_1 + V_2) = (V_1 - V_2) \cdot 2 \cdot V = \left(\frac{V}{1-a} - V_1 \cdot (1-2a) \right) \cdot 2 \cdot V$$

$$V_1^2 - V_2^2 = \left(\frac{V}{1-a} - \frac{V}{1-a} \cdot (1-2a) \right) \cdot 2 \cdot V = \frac{V}{1-a} \cdot (1-(1-2a)) \cdot 2 \cdot V$$

$$V_1^2 - V_2^2 = \frac{4 \cdot a \cdot V^2}{1-a}$$

Replacing the identity of the axial force we get:

$$\frac{a}{1-a} = \frac{l \cdot h \cdot z}{8\pi \cdot A} \cdot \int_0^{2\pi} \frac{W^2}{V^2} \cdot (C_r \cdot \text{sen}\theta - C_t \cdot \cos\theta) \cdot d\theta \qquad (5.2)$$

Remember that, by definition, the solidity of the rotor, σ, is given by:

$$\sigma = \frac{l \cdot h \cdot z}{A}$$

Furthermore, W^2 is known since the beginning, repeated here for convenience:

$$W^2 = (V \cdot \text{sen}\theta)^2 + (\Omega \cdot R + V \cdot \cos\theta)^2$$

Hence, the quotient between the squares of W and V becomes:

$$\frac{W^2}{V^2} = (\text{sen}\theta)^2 + \frac{(\Omega \cdot R)^2}{V^2} + 2 \cdot \frac{\Omega \cdot R}{V} \cdot \cos(\theta) + (\cos\theta)^2$$

$$\frac{W^2}{V^2} = 1 + \frac{(\Omega \cdot R)^2}{V^2} + 2 \cdot \frac{\Omega \cdot R}{V} \cdot \cos(\theta)$$

By analogy with the specific speed, we can define the coefficient λ' as:

$$\lambda' = \frac{\Omega \cdot R}{V}$$

Hence, Equation 5.2 can be rewritten as follows:

$$\frac{a}{(1-a)} = \frac{\sigma}{8\pi} \cdot \int_0^{2\pi} (1 + \lambda'^2 + 2 \cdot \lambda' \cdot \cos\theta) \cdot (C_r \cdot \text{sen}\theta - C_t \cdot \cos\theta) \cdot d\theta \qquad (5.3)$$

In the same way of *Glauert's* integral, Equation 5.3 can be calculated numerically for discrete values of σ and λ'.

Having calculated the integral, the value of a is calculated directly, and with it we can finally calculate the forces on the blade for each speed of wind and specific speed λ'.

Since, by definition:

$$\lambda = \frac{\Omega \cdot R}{V_1} \text{ and } V = V_1 \cdot (1-a)$$

We can then deduct the relationship between λ and λ':

$$\lambda' = \frac{\lambda}{1-a}$$

In Table 5.1, the reader can see an example of the results obtained from Equation 5.3, calculated in the interval $0 < \theta < 45°$, assuming $\lambda' = 10$ and $\sigma = 13\%$ and finally deducing the values of C_t and C_r from the lift and drag table of an airfoil NACA 0018. From the discrete integral from 0° to 180°, it is possible to calculate the values of λ and of a.

Consequently, utilizing the relation between λ and λ' deduced above, we can transform Table 5.1 to Table 5.2, from which in turn we will be able to easily calculate the forces, torque, and power, as a function of the values of V_1 and λ.

TABLE 5.1

Using a Spreadsheet for Solving Numerically Equation 5.3, Assuming $V = 1$

Input data

λ'	10
σ	13%
Airfoil	NACA 0018 (data from Sheldahl, Re = 160,000)

Calculated variables

A	0.488
λ	5.120

θ(rad)	$\theta°$	V_x	$V_t + V_x$	V_y	α	$\alpha°$	C_t	C_r	Discrete Integral
0	0	1	11	0	0.00000	0.0	–0.0128	0.0000	1.5488	1.5453679
0.087266	5	0.9961947	10.996195	0.0871557	0.00793	0.5	–0.0128	0.0000	1.541936	2.0836224
0.174533	10	0.9848078	10.984808	0.1736482	0.01581	0.9	–0.0124	0.0551	2.625309	3.6665794
0.261799	15	0.9659258	10.965926	0.258819	0.02360	1.4	–0.0110	0.1102	4.70785	6.2282663
0.349066	20	0.9396926	10.939693	0.3420201	0.03125	1.8	–0.0087	0.1653	7.748683	9.7126822
0.436332	25	0.9063078	10.906308	0.4226183	0.03873	2.2	–0.0054	0.2203	11.67668	13.764175
0.523599	30	0.8660254	10.866025	0.5	0.04598	2.6	–0.0019	0.2647	15.85167	18.211953
0.610865	35	0.819152	10.819152	0.5735764	0.05297	3.0	0.0025	0.3091	20.57224	21.731035
0.698132	40	0.7660444	10.766044	0.6427876	0.05963	3.4	0.0025	0.3091	22.88983	25.787729
0.785398	45	0.7071068	10.707107	0.7071068	0.06595	3.8	0.0080	0.3603	28.68563	31.759297

TABLE 5.2

Results of the Calculations Performed with Table 5.2, with Different Values of V and λ', Defined Arbitrarily

λ	$\sigma = 13\%$	$\sigma = 12\%$	$\sigma = 8\%$
	C_P	C_P	C_P
2			
3	0.14		
3.5	0.2	0.19	0.13
4	0.21	0.22	0.19
4.5	0.18	0.15	0.2
5	0.14	0.05	0.16
6	0	0	0.09

5.3 Design of H-Type Darrieus Turbines

The calculation procedure of Darrieus turbines is analogous to the one explained for horizontal axis turbines: starting from the pre-dimensioning, then choosing an airfoil based on given criteria, calculating the values of a for different solidities and specific speed, and finally calculating P and C_P with the formulas of the forces acting on the blade.

5.3.1 Pre-dimensioning of H-Type Darrieus Turbines

The following procedure is valid only for H-type, which is the simplest alternative to design and build. The designer must define the desired power and the nominal speed of the wind according to the place, the area exposed to wind results from the general equation of the wind turbine's power, assuming $C_P = 0.30$ (a bit optimistic for small size turbines) and $\rho = 1.21\,\mathrm{kg/m^3}$, hence, grouping the constants, we get:

$$P = 0.18 \cdot D \cdot h \cdot V^3 \rightarrow D \cdot h = \frac{P}{0.18 \cdot V^3}$$

where:
P = power (W)
D = diameter (m)
h = height of the rotor (m)
V = speed of the wind assumed for the site (m/s)

The ratio between D and h is arbitrarily defined by the designer, but for aesthetical reasons, usually $h > D$. The ratio $h/D = 1.4142$ (the same ratio between height and width of an A4 sheet) provides a pleasant aspect for human eyes. The number of blades, z, of the Darrieus turbines can be only 2 or 3. The 2-bladed turbines present some problems of vibrations and difficulties for their dynamic balancing; hence, 3-bladed rotors should be preferred. Rotors with four or more blades, because of their higher solidity and conspicuous aerodynamic interferences between the blades upstream and those downstream the shaft, tend to increase the C_P, but at the same time to narrow the power curve, which will assume the typical "spike" shape. For such reason, 4-bladed rotors are seldom employed.

Once V, h, and D, are known, we define arbitrarily a value of σ in the range 8%–15%. With the method described until now, the value 13% provides the most reliable results. Once all the said design parameters are known, the chord of the airfoil, l, is automatically defined.

Before proceeding to calculate the performance of the rotor, it is necessary to choose an airfoil. This argument requires a treatise on its own, presented in the next point.

5.3.2 Choosing the Airfoil for the Rotor

The same considerations already exposed in Chapter 4 for horizontal axis wind turbines are valid for the design of Darrieus turbines, with a small variant: symmetric profiles are preferable for Darrieus turbines, because the angle of incidence is negative along a part of the trajectory. In the eggbeater-type turbines, the shape of the blades approximates a catenary, hence the material of the blades works mainly under pure traction, and hence slender profiles, featuring better aerodynamic performance, are the best option. The curve that ensures pure traction loads along the blade, called a *troposkien* (Blackwell, 1974), has quite a complex mathematical expression and is difficult to build; hence, it will not be discussed in this book. For the construction of small turbines, it is easier to build straight blades, disposed as an H. Since in this type of rotor the loads on the blades will be of flexion, it is necessary to employ thicker profiles, in order to grant enough rigidity and resistant section. Profiles like NACA 0015 or NACA 0018 were employed largely in the past. More modern and efficient versions, like the Eppler 520 series, or the NACA 63_2-015, should be preferred. According to *Migliore* and *Fritschen*, the latter would yield 15% more efficiency compared to NACA 0015. In order to analyze the performance when $\lambda < 3$, it is necessary to have data of C_z and C_x on the whole range of incidence angles: from 0° to 180° for symmetric airfoils, and from 0° to 360° for asymmetric ones. In the example shown in Tables 5.1 and 5.2, the Author adopted data of NACA 0018, both because these are available from the literature (*Sheldahl*), and in order to compare the data calculated with the spreadsheet with the experimental data, measured in the wind tunnel by *Sabaeifard*, *Razzaghi*, and *Forouzandeh*.

5.3.3 Calculating the Coefficient *a* for Different Arbitrary Values of λ' and Determination of the Forces, Torque, C_P, and λ

Having already chosen the airfoil, defined the dimensions of the rotor and the dimensions of the blades, we can modify Table 5.1 in order to calculate the forces on the blades, and hence the torque and power, for arbitrary values of λ' and V. Once the power is known, the numerical integration of Equation 5.3 allows us to calculate also the coefficient *a*, from which we then calculate the values of Ω, λ, and V_1 corresponding to the values of λ' and V arbitrarily defined and, in final step, the C_P of the turbine at such speeds. The sequence of calculation involves defining arbitrarily a value of σ in the range of 10%–20%. Remember that, by definition:

$$\sigma = \frac{z \cdot l \cdot h}{D \cdot h}$$

Therefore, the chord of the airfoil is automatically defined:

$$l = \frac{\sigma \cdot D}{z}$$

Once l, z, and h are known, divide the circular path of the blade in 72 or more points, and calculate the apparent wind W and the angle of attack α in each point. The speed V and the specific speed λ' are arbitrarily defined by the designer. It is then possible to calculate for each point of the circular path the forces F_t and F_n, and hence the torque, the power, and the maximum C_P.

The nominal speed of rotation, N, results then automatically from the definition of λ (Section 2.3.3.2), hence:

$$\lambda = \frac{\Omega \cdot R}{V} = \frac{\pi \cdot D \cdot N}{60 \cdot V} \rightarrow N = \frac{\lambda \cdot 60 \cdot V}{\pi \cdot D}$$

where:
Ω = angular speed (rad/s)
N = angular speed (R.P.M.)
V = speed of the wind (m/s)
$R = D/2$ = radius (m)

Now the power P is known, and having calculated the coefficient a thanks to the numerical resolution of the integral in Equation 5.3, it is possible to obtain the values of V_1 and λ, corresponding to the values of V and λ' arbitrarily imposed in the first step, and hence the C_P for each operational condition.

The whole procedure is summarized in the file *Darrieus-H.xlsx* provided with this book. Its use will be explained in Section 5.6. Practical Exercises.

The values of V_1, λ, and C_P obtained with the spreadsheet can be tabulated, as shown in Table 5.2.

The experience shows that, for low values of solidity (from 8% to 11%), the range of specific speeds within which the turbine operates with good efficiency is wider (hence, a flat curve, in general in the range $4 < \lambda < 7$), but the maximum C_P of the turbine remains below 0.30. On the contrary, increasing the solidity (from 12% to 20%) the optimum operation of the turbine converges to a single point, where C_P is very close or slightly higher than 0.30, but the operation range of the rotor is restricted to $3 < \lambda < 6$. A good compromise between an acceptable value of maximum C_P and operational range width is usually obtained with $12\% < \sigma < 15\%$.

Figure 5.5 shows graphically the comparison between the values of C_P and of λ, obtained with the spreadsheet *Darrieus-H.xlsx* and the experimental values obtained in the wind tunnel by *Sabaeifard, Razzaghi, and Forouzandeh*. These built a small *Darrieus* turbine having straight blades of $40 \times 80\,cm^2$. Observe that the numerical results fit well enough the experimental ones, although resulting systematically inferior. The reasons are multiple: first, the method of calculation adopted here is the simplest existing in the literature, and second, we assumed that the coefficients C_z and C_x of the airfoil would have the same values both in dynamic and static conditions. Such supposition is not completely true because the experience shows that, causing the angle of attack of the airflow to vary

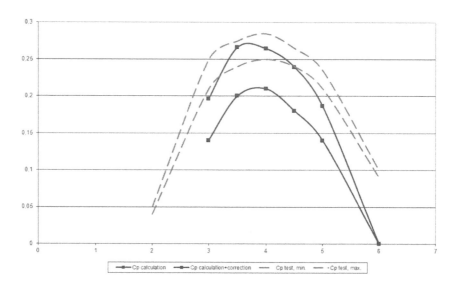

FIGURE 5.5
Comparison between the curve of C_p as a function of λ obtained with the spreadsheet of Table 5.2, with and without the correction for the *Katzmayr* effect and the curves measured in the wind tunnel on a model rotor. In both cases, the airfoil was NACA 0018, working with 100,000 < Re < 160,000. Experimental data from *Sabaeifard, Razzaghi,* and *Forouzandeh*; graphic elaboration by the Author.

periodically on a symmetric airfoil—as in the case of the blades of the Darrieus turbines—then the average values of drag tend to decrease, and in some cases, they can become net forward forces. This last paradox is better known in the literature as *Katzmayr effect*, from the name of the researcher that described it for the first time in 1922. Our extremely simplified model does not take into account the said effect, and consequently tends to underestimate by nearly 20% the real performance of the turbine. In any case, underestimating the power for a given size is not a problem, because we must consider that small-sized generators and inverters have low overall efficiencies, nearly 90%, and that with time the blades will get dirty and lose their aerodynamic efficiency. For practical purposes, the design performed with our simplified method will fit relatively well the real energy production in the field along the life of the turbine. From tests performed by *R.H. Wickens* in Canada in 1985, it is known that airfoil NACA 0018, under cyclical variations of the angle of attack up to 30°, tends to delay by 5° its stall angle. At the same time, the coefficient C_r rises by 20% and the coefficient C_t varies from a minimum of 10% up to a maximum of 40%. Consequently, C_p grows, demonstrating that if our C_p values had been calculated with the correct dynamic coefficients instead of the static ones, they would fit well with the tests in wind tunnel carried out by

Sabaeifard et al., at least in the interval $3 < \lambda < 5$, which is the only interesting one for practical purposes.

5.4 Analysis of the Aerodynamic Features and Constructive Choices of the Rotor

5.4.1 Speed Control

By definition, Darrieus turbines are fixed pitch turbines. The rotation speed and the output power can be controlled only by passive or active stall. In the first case, the rotation speed of the turbine is not limited and, as soon as $\lambda > 8$, the mechanical power output at the shaft falls drastically until reaching an equilibrium with the power absorbed by the alternator. Under extreme winds, the alternator is disconnected and the turbine is left turning idle at high speed, but the load on the structure is anyway a fraction of the load under normal operation because the energy absorbed from the wind is very small, just the friction at the bearings and the aerodynamic drag of the rotor.

In the second case, the rotor is slowed down by means of any device that dissipates the excess of energy that the alternator cannot covert. Such devices can be, for example, auxiliary resistances; aerodynamic breaks driven by centrifugal devices, *flaps* or *spoilers* inducing the stall even at low angles of attack, or finally, mechanical breaks controlled electronically. Whatever the breaking method, its scope is to slow down the rotation speed of the turbine to the point in which $\lambda < 4$, making the driving torque null or negative, and the rotor will not be able to spin any more.

The determination of the power curve—as shown in former Section 5.3.3—can be carried out easily with a spreadsheet, where we will vary the values of V_1 and λ until finding the equilibrium condition between the power at the shaft and the power absorbed by the generator, obtaining then the rotation speed N. Finally, we will be able to plot the curve of P, as a function of N, or of P as a function of V_1, which in turn will allow plotting the curve of C_P as a function of λ explained before.

5.4.2 Production of the Blades

The same considerations explained in Section 4.4 for the horizontal axis wind turbines are valid also for H-type Darrieus turbines. The use of extruded aluminum profiles or of plywood boards is relatively easy for the artisan. It is, furthermore, possible to employ aluminum sheet shaped around a tube of the same material, adequately riveted and filled with polyurethane foam.

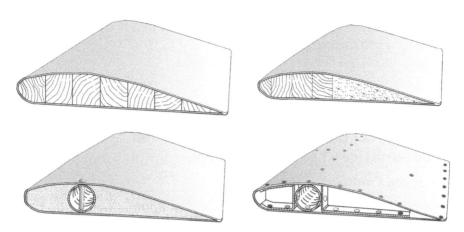

FIGURE 5.6
Some constructive solutions for the manufacture of the blades. In clockwise sense, from top left: Blade in plywood with plastic resin sheath, blade in plywood-foam composite with plastic resin sheath, blade in aluminum foil riveted to a metallic ribs, blade of fiberglass reinforced resin, filled with foam and central aluminum tube. With thanks to prof. Arch. Mario Barbaro for digitizing the Author's sketches.

The Figure 5.6 illustrates some of the possible constructive solutions for H-type Darrieus rotors.

5.5 Practical Example: Making a Low-Cost Darrieus Rotor

5.5.1 Definition of the Problem

Given the same situation described in Section 4.5.1, the goal is to make by hand a low-cost H-Type Darrieus turbine for the basic supply of energy in rural areas in a developing country. An alternator with nominal rating 12 V × 60 A, recycled from dismantled car, will be the energy converter. The nominal rotation speed of the alternator is estimated in 1200 R.P.M. From an anemometric study of the place, the maximum annual energy corresponds to a speed of the wind equal to 5.5 m/s, hence, the turbine must reach its nominal power at said speed.

5.5.2 Pre-dimensioning

Applying the formula presented in Section 5.3.1 we get:

$$D \cdot h = \frac{P}{0.18 \cdot V^3} \rightarrow D \cdot h = \frac{60 \cdot 12}{0.18 \cdot 5.5^3} = 24 \, \text{m}^2$$

Assume arbitrarily $h/D = 1.4142$ just for aesthetical reasons, hence:

$$D^2 = \frac{24}{1.4142} \rightarrow D = 4.12\,\text{m} \rightarrow h = 5.83\,\text{m}$$

The rotation speed is calculated assuming $\lambda = 4$, hence:

$$N = \frac{\lambda \cdot 60 \cdot V}{\pi \cdot D}$$

$$N = \frac{4 \cdot 60 \cdot 5.5}{\pi \cdot 4.12} = 102\,\text{R.P.M.}$$

As in the example 4.5.1, it is necessary to insert a multiplication device between the shaft of the turbine and the alternator. The multiplication ratio is:

$$Z = \frac{N_2}{N_1} = \frac{1,200}{102} = 11.8$$

The speed multiplication can be done with a pulley and belt system, friction wheels, pinion and chain, or gearboxes. For small power (about 900 W at the shaft of the turbine) the most adequate solutions for our case could be a system of pulleys and belts (recycled from old vehicles) or friction wheels made of plywood rimmed with rubber.

5.5.3 Selection of the Airfoil

A low-cost turbine is desired, and only basic tooling is available. The most adequate alternatives are plywood boards, or curved aluminum foil. For this example, we will assume making the blades with plywood boards, adequately profiled. The easiest airfoil for handmade construction is Clark Y, whose aerodynamic features are shown in Chapter 12. Of course, it could be possible to produce symmetric profiles, but that would mean double handwork, i.e., streamlining two surfaces per blade instead of just one.

5.5.4 Dimensioning of the Blade

From the considerations presented in Section 5.3.3, we define arbitrarily

$$\sigma = 13\% \text{ and } z = 3$$

Consequently, the chord of the airfoil of each blade results from the following relationship:

$$l = \frac{\sigma \cdot D}{z}$$

$$l = \frac{0.13 \cdot 4.12}{3} = 0.18\,\text{m}$$

In order to calculate the approximate curve of aerodynamic performance when employing the asymmetric Clark Y airfoil instead of the usual symmetric airfoils, make a copy of the file Darrieus-H.xlsx and replace the values of C_z and C_x in the sheet NACA 0018. Also try different pitch angles between the chord and the tangent to the circular path. It is necessary to convert the values C_z and C_x presented in the lift and drag table of Clark Y airfoil (Chapter 12) in the corresponding values of C_t and C_r, for each angle α, employing the spreadsheet and the already known formulas:

$$C_t = C_z \cdot \text{sen}\,\alpha - C_x \cdot \cos\alpha$$

$$C_r = C_z \cdot \cos\alpha + C_x \cdot \text{sen}\,\alpha$$

5.6 Exercise

Suppose building the same turbine already pre-dimensioned in the former Section 5.5, but employing an E169 symmetric airfoil. Such profile was specially conceived for low-Re applications. Assume Re = 200,000 and refer to Section 12.3.6. for the aerodynamic data. What will the minimum λ be, assuming the airfoil does not stall in any point of its path?

Solution
From the cited table, the maximum value of α at which the profile is still within the separation limit is ±17.75°. Now observe Figure 5.7. The angle α

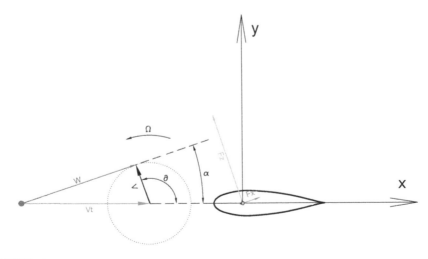

FIGURE 5.7
Condition of maximum pitch angle.

will reach 17.75° when the vector W is tangent to the circle described by the rotating vector V.

By definition:

$$\lambda' = \frac{\Omega \cdot R}{V} = \frac{V_t}{V}$$

By construction (Figure 5.7):

$$\sec(17.75°) = \frac{V_t}{V} = \lambda'$$

$$\lambda' = 3.280$$

In order to calculate the minimum value of λ for a stable rotation, it is necessary to know the value of a at such limited operational condition. In order to calculate a, the discrete integral must be calculated. Since

$$\lambda' = \frac{\lambda}{1-a}$$

and $(1 - a)$ is always smaller than 1, we can estimate that λ will be a value between 2 and 3. The reader is invited to make another copy of the file *Darrieus-H.xlsx* and to replace the data of the NACA 0018 by those of the Eppler E169 airfoil. The resulting value of λ should be 2.677 or very close to it (depending on rounding errors).

We can derive an important conclusion from this example, which in general is valid for all kinds of aerodynamic action turbines: when designing a small turbine that has to work at low Re, it is preferable to search for airfoils especially suitable for such condition, like the E169. The reason is that such airfoils stall at higher pitch angles, allowing then a wider operational range of λ. Profiles of the NACA 00XX series are suitable for Re > 1,000,000, hence for bigger turbines and/or stronger winds.

Bibliography

Bedon G., Raciti Castelli M., and Benini E., *Evaluation of the Effect of Rotor Solidity on the Performance of an H-Darrieus Turbines Adopting a Blade Element-Momentum Algorithm*, World Academy of Science, Engineering and Technology, volume 6, 2012.

Blackwell E., and Reis B., *Blade Shape for a Troposkien-Type Vertical Axis Wind Turbines*, Sandia Laboratories Energy Report SLA-74-0154, Albuquerque, USA, 1974.

Le Gourière D., *L'Énergie Éolienne – Théorie, conception et calcul practique des installations*, deuxième édition, Eyrolles, Paris, 1982.

Migliore P., and Fitschen J., *Darrieus Wind Turbines Airfoil Configurations - A Subcontract Report*, SERI/TR-11045-1, UC Category: 60, Solar Energy Research Institute, Colorado, USA, 1982.

Nasolini R., *Studio e ottimizzazione dell'avviamento di un generatore ad asse verticale*, degree thesis in mechanical engineering, University of Bologna, 2010–2011.

Ober S., *Note on the Katzmayr Effect on Airfoil Drag*, NACA Report, volume 214 technical note, 1925.

Sabaeifard P., Razzaghi H., and Forouzandeh A., *Determination of Vertical Axis Wind Turbines Optimal Configuration through CFD Simulations*, International Conference on Future Environment and Energy, IPCBEE volume 28, IACSIT Press, Singapore, 2012.

Sheldahl R., and Klimas P., *Aerodynamic Characteristics of Seven Symmetrical Airfoil Sections for Use in Aerodynamic Analysis of Vertical Axis Wind Turbines*, Sandia National Laboratories Energy Report SAND 80-2114, Albuquerque, USA, 1981. Available for free download from the following URL: http://prod.sandia.gov/techlib/access-control.cgi/1980/802114.pdf.

Wickens R.H., *Wind Tunnel Investigation of Dynamic Stall of a NACA 0018 Airfoil Oscillating in Pitch*, National Research Council Canada, Aeronautical Note NAE-AN-27, NRC no 24262, 1985.

Wilson R., and Lissaman P., *Applied Aerodynamics of Wind Power Machines*, Oregon State University, 1974.

6

Practical Design of Savonius Turbines and Derived Models

6.1 Generalities

Vertical axis rotors based on the principle of differential action have been known since ancient times. The Finnish engineer *Sigurd Savonius* gave his name to the rotor he "invented" in 1921 and patented eight years later. In this chapter, we are going to analyze what improvements *Savonius* added to the rotors already existent in Northern Europe since the beginning of the 20th century. Most importantly, we will analyze, in general, the working principle of the turbines belonging to the family named scientifically "pannemones" (from Ancient Greek παν = all and πνεὖμων = air, wind, i.e., "able to turn on any wind") and commonly referred to as pinwheels. For the sake of truth, the vertical axis turbines called "pannemones" by ancient Greeks are a Persian invention featuring clapping screens that become alternatively perpendicular or parallel to the wind. The fundamental feature of pinwheels is the ability to rotate regardless of the direction of the wind. The physical principle behind their operation is the difference between the drag force on a concave body, and that on a convex body presenting the same cross-section to the wind.

Now let us consider the simplest type of pinwheel, usually employed for the production of cup anemometers. Figure 6.1 shows the cross-section of such device.

The general expression of the drag force exerted by a fluid stream on a body is:

$$F = \frac{\rho}{2} \cdot S \cdot V^2 \cdot C$$

Where:

ρ = density of air (kg/m³)
V = speed of the wind (m/s)
S = cross-section of the body, perpendicular to the wind (m²)
C = non-dimensional coefficient, depending on the body's shape

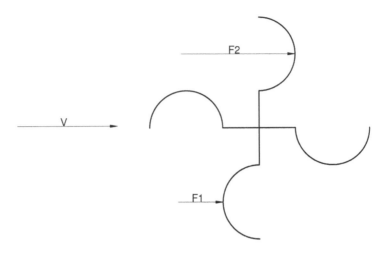

FIGURE 6.1
Section of a cup anemometer.

The value of C corresponding to the convex side of the semisphere is 0.34. The correspondent value of the concave part is 1.33; hence, $F_2 > F_1$ and a net torque with respect to the turning axis will appear, which causes the turbine to rotate.

Suppose that the center of the cups moves with tangential speed V_t under a wind of speed V. The relative speed between the wind and the cups presenting their concave part to this latter, is the difference between V and V_t. The relative speed between the wind and the cups presenting their convex part to it is the sum of V and V_t. The forces on both sides will be proportional to the square of the sum $(V + V_t)$ and of the difference $(V - V_t)$.

The power can be calculated as follows:

$$P = M \cdot \Omega = (F_1 - F_2) \cdot R \cdot \Omega$$

Note that $\Omega \cdot R = V_t$

Hence,

$$P = \frac{\rho}{2} \cdot S \cdot \left[C_1 \cdot (V - V_t)^2 - C_2 \cdot (V + V_t)^2 \right] \cdot V_t$$

It is possible to demonstrate that the power is maximum, when

$$V_t = \frac{2 \cdot k \cdot V - V \cdot \sqrt{4 \cdot S^2 - 3 \cdot j^2}}{3 \cdot j}$$

Where:

$$k = C_1 + C_2$$
$$j = C_1 - C_2$$

Applying the definition of λ, the maximum power is attained when:

$$\lambda = \frac{V_t}{V} = \frac{2 \cdot k - \sqrt{4 \cdot S^2 - 3 \cdot j^2}}{3 \cdot j}$$

In general, the efficiency of practically feasible panemones is maximum for λ values in the range 0.3–0.6.

The *Savonius* rotor is not a pure differential action turbine, like the pinwheel. The modification to the classical pinwheel introduced by *Savonius* involves a certain overlapping of the two semi-cylinders, *e*, which deflects part of the airflow (Figure 6.2). The airflow around the convex side of the semi-cylinder produces some supplementary aerodynamic forces, improving the overall efficiency.

Endless tests have been carried out by many researchers in order to find out the optimum distance, *e*, between the inner edges of the semi-cylinders. Figure 6.3 shows the C_p measured on five different models of *Savonius* rotor, as a function of λ, and the start torque as a function of the wind direction, in polar coordinates. Table 6.1 provides the geometric features of each rotor.

Figure 6.4 shows some areas where the start torque is negative. The way to avoid such inconvenience is to stack two identical rotors, with 90° of pitch between both main transversal axes. With such spatial disposition, the curves of C_p as a function of λ do not change, but the start torque becomes always positive. Figure 6.5 shows such solution. In "low cost" models, the rotors are usually cut out of old industrial steel drums, usually employed for transporting liquids. The solution shown in Figure 6.5 makes the torque in normal operation more regular. Indeed, a simple straight rotor features pulsating torque, with the consequent problems of vibrations transmitted to the structure and lower lifespan.

The original rotor designed by *Savonius* corresponds to rotor IV in Figure 6.3, but studies in wind tunnel demonstrate that the best solution is rotor II.

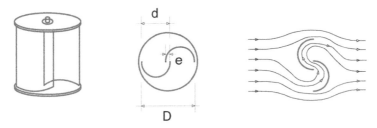

FIGURE 6.2
Side view, cross-section, and airflow across a *Savonius* rotor.

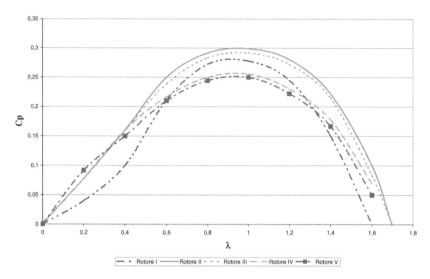

FIGURE 6.3
Differences between the aerodynamic performances of the *Savonius* rotor as a function of the ratio *e/D*, according to *Newman* and *Lak Ah Chai*, published by *Le Gourière*. Graphical elaboration by the Author.

TABLE 6.1

Geometric Features of Each of the Rotors Shown in Figure 6.3

Rotor	e/D Ratio
I	0
II	1/6
III	1/4
IV	1/3
V	0.43

Remember that the power of a wind turbine is:

$$P = \frac{\rho}{2} \cdot C_p \cdot S \cdot V^3$$

Replacing the values of C_p obtained with the experiments in wind tunnel (maximum) for rotors II and IV, assuming $\rho = 1.25\,\text{kg/m}^3$, we obtain:

$P = 0.18 \cdot S \cdot V^3$ For the rotor model II

$P = 0.15 \cdot S \cdot V^3$ For the original rotor of *Savonius*, model IV

Where:
S = area exposed to wind = $h \cdot D$ (m²)

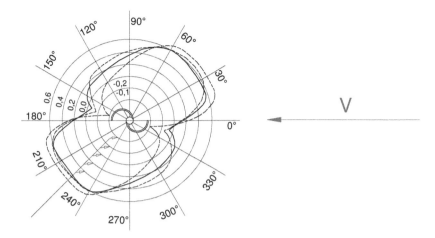

FIGURE 6.4
Start torque, C_a (Nm), as a function of the wind direction. The continuous line corresponds to rotor type II. Data from *Le Gourière*, drawings by the author.

The power will be maximum when $0.9 < \lambda < 1$ (experimental results). The torque under such conditions can be calculated as a function of the torque coefficient, C_m, from the following relationship:

$$C_p = C_m \cdot \lambda$$

from which it is possible to deduce that:

$$M = \frac{\rho}{2} \cdot C_m \cdot R \cdot S \cdot V^2$$

Where:

$$R = d - 0.5 \cdot e$$

The variation of C_m with λ is shown in Figure 6.6.

Other shapes of semi-cylinders have been experimented apart from circular, parabolic, hyperbolic, etc., but the best performance is still that of the rotor-type II.

A construction variant is adding a screen, which is oriented by an air rudder in order to keep the lowest drag force on the inactive semi-cylinder. In practice, such solution is seldom applied because it complicates the turbine and increases the manufacturing costs, providing a maximum increase of the C_p value of only 38% in the best case (Altan et al.).

Figure 6.7 shows the screen that provides the best results, according to Altan et al.

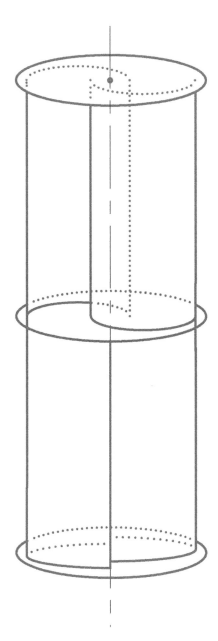

FIGURE 6.5
Rotors with 90° pitch, useful to avoid the areas of negative start torque and to make the torque in normal operation more regular.

Tests on *Savonius* rotors with three semi-cylinders, shown in Figure 6.8, demonstrate that the triple semi-cylindrical geometry provides a more regular torque under steady state operation, though this latter is still pulsating.

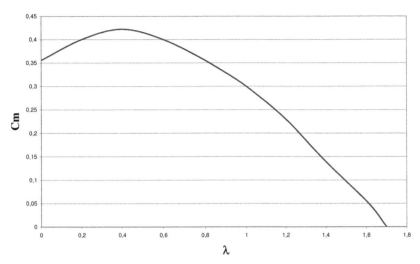

FIGURE 6.6
Variation of C_m as a function of λ for rotor model II, according to the experiments of *Newman* and *Lak Ah Chai* published by *Le Gourière*. Graphic elaboration by the Author.

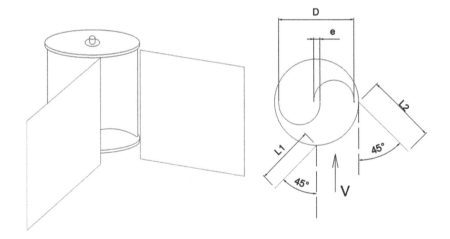

FIGURE 6.7
Screened *Savonius* rotor, data according to Altan et al., sketch by the Author. The best results of several combinations were obtained with $\alpha = \beta = 45°$; $l_1 = 1.4$ D; $l_2 = 1.62$ D; $e/D = 0.15$, as shown in the picture. Such solution requires some sort of vane (not shown) for keeping the screens oriented to the wind's direction.

The best results, according to *Morshed*, vary with the Reynolds number, *Re*, while the *e/D* ratio seems to become a second-order parameter compared to *Re*. *Morshed's* study is limited to three rotors, having $e/D = 0$ (semi cylinders joined at the center, i.e., a classical pinwheel), $e/D = 0.12$, and $e/D = 0.26$. For practical purposes, we can conclude that the *e/D ratio* in three-semi cylinder

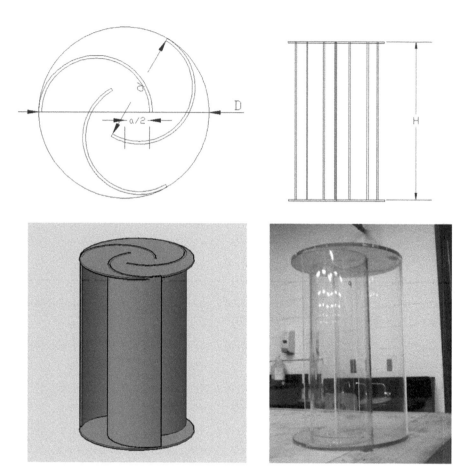

FIGURE 6.8
One of the three variants of *Savonius* rotors, having three semi-cylinders, tested by *Morshed*.
From the top, in clockwise sense: cross and length sections, 3D view, and model in Plexiglas
tested in the wind tunnel.

Savonius rotors, is not a critical design parameter for the design and the C_p is
not very different from that of the original *Savonius* model. The advantage of
the three semi-cylinder solution is the absence of zones with negative start
torque. Its disadvantage includes higher material cost and weight. Indeed,
comparing a three semi-cylinder rotor with the one shown in Figure 6.5,
this latter would provide double exposed area, hence double the power, with
only 30% more material.

   Another variant, developed with the scope of regularizing the torque, is
the twisted rotor, like the one produced by some commercial manufacturers,
for instance, *Helix Wind* (A company failed in 2012, http://news.nationalgeo-
graphic.com/news/energy/2012/08/120820-helix-wind-collapse/). It is essen-
tially a *Savonius* rotor having helicoidal semi-cylinders instead of straight

ones. Each helix describes 90° from its base to its top. With such geometry, the resulting start torque is more uniform, and the look more appealing than those of the classic *Savonius* rotor with straight semi-cylinders. On the other hand, the overall performance is probably lower (no data available, since the company failed) and its construction is for sure more complicated.

Savonius rotors are suitable for their construction and installation in zones with weak winds (because of their high torque with slow rotation speed) and/or in developing countries or disaster zones (because of their production simplicity and the possibility of recycling low cost, widely available components, like dismissed metal or plastic drums).

Their high start and normal operation torques make *Savonius* rotors particularly suitable for driving piston, peristaltic, and diaphragm pumps. If the scope is generating electricity, horizontal axis rotors should be preferred, because their higher specific speed makes them easier to adapt, either by direct drive or coupling with low-ratio gearboxes. Nevertheless, if an adequate speed multiplication and/or an alternator with a high number of poles are available, it is possible to employ also a *Savonius* rotor for the generation of electricity. The manufacturing cost will increase and the overall efficiency will drop, because the gear or pulley multiplication introduces additional friction. The main disadvantage of the *Savonius* rotor is that, for a given nominal output power and wind speed, its size and weight will be bigger than those of aerodynamic action turbines, because the C_p of the first is less than half that of the latter.

6.2 Practical Calculation of *Savonius* Rotors

In general, the start point for the calculations is the size of the available semi-cylinders, because these are usually cut out of commercial drums employed for the transport of liquids. Let us follow systematically the design phases.

6.2.1 Determine the Power Obtainable from a Given Speed of Wind

Knowing the dimensions and the number of available drums apply the formula already presented in Section 6.1:

$$P = 0.18 \cdot S \cdot V^3$$

Remember that said formula is valid for the rotor type II, where $e/D = 1/6$ and hence, the geometry of the rotor is automatically defined. If at least two drums are available, the best solution is to stack two rotors, pitched 90°, as shown in Figure 6.5.

Example: two steel drums are available, having $d = 0.7$ m, $h = 1$ m. Cutting them lengthwise, their optimum disposition implies that the total diameter of the rotor, D, will be:

$$D = 2d - e = 2d - \frac{D}{6} \rightarrow \frac{7}{6}D = 2d \rightarrow D = \frac{12}{7}d$$

D is hence 1.2 m. Since both rotors will be stacked with 90° pitch, the total height of the rotor will be 2 m and hence, $S = 2.4$ m².

6.2.2 Determine the Torque

Knowing the maximum C_p, the relationship with λ is useful for determining C_m, according to the formula already introduced in Section 6.1:

$$C_p = C_m \cdot \lambda$$

and

$$M = \frac{\rho}{2} \cdot C_m \cdot R \cdot S \cdot V^2$$

6.2.3 Determine the Necessary Torque for Driving the Pump

Knowing the maximum torque that the rotor is able to produce, and the torque required by the pump under nominal operation, the multiplication ratio of the gearbox, k, is automatically defined. Another option is to impose arbitrarily a value of k and, in base of it, searching the most suitable pump in the catalogs of several manufacturers.

The power absorbed by the pump results from the following formula:

$$P = \frac{\rho \cdot k \cdot N \cdot q \cdot H}{\eta}$$

Where:
ρ = specific weight of water = 9.800 N/m³
k = multiplication ratio
N = speed of the rotor (obtained assuming the maximum efficiency at the nominal wind speed $\lambda = 1$) (turns/s)
q = volume of water pumped at each turn of the pump's shaft (m³)
H = equivalent head (manometric + friction losses) (m)
η = pumping efficiency (between 0.5 and 0.78, depending on the pump and the multiplication system employed)

Equalling the pump's power to the rotor's power, we can then determine the stroke of the pump q (m³) and search a suitable model in a catalog.

Otherwise, if the stroke is already known, because the pump is already available or because there is no better choice, then calculate N and check if it coincides with the speed of rotation required by the pump. If N does not coincide with the nominal speed of the pump, recalculate a new value of λ, determine the rotor's power for it, and finally compare it with the power required for pumping. Repeat the said process by successive approximations until finding the optimum working point of the rotor.

6.2.4 Calculation of the Mass Flow

Once all the design parameters are known, as explained above, we are able to calculate the nominal mass flow:

$$Q = k \cdot N \cdot q$$

6.2.5 Curve of Mass Flow as a Function of the Wind Speed, V

Repeating the full process described above with diverse wind speeds, we will have the necessary data for plotting the curve of mass flow as a function of V.

Bibliography

Altan B.D., Atilgan M., and Ozdamar A., An experimental study on improvement of a Savonius rotor performance with curtaining, *Experimental Thermal and Fluid Science* 32, 1673–1678, 2008.

Le Gourière D., *L'Énergie Éolienne – Théorie, conception et calcul practique des installations*, deuxième édition, Eyrolles, Paris, 1982.

Morshed K.N., *Experimental and Numerical Investigations on Aerodynamic Characteristics of Savonius Wind Turbine with Various Overlap Ratios*, Electronic Theses & Dissertations. 773. 2010, http://digitalcommons.georgiasouthern.edu/etd/773.

Rosato M., *Diseño de máquinas eólicas de pequeña potencia*, Editorial Progensa, Sevilla, 1992.

7

Engineering of the Support Structures for Wind Turbines

7.1 Generalities

The design of the most delicate part of a small wind power installation, its support structure, requires no special algorithms or models: the classical formulas and methods of civil engineering are enough for a safe design. It is important to respect all relevant engineering and construction codes, in order to grant the structural integrity under the worst operational conditions. It is possible to build the support tower of a wind turbine with a large variety of materials and shapes. The most suitable solution for a given case will depend on the turbine's power and the locally available resources. Simple wooden, steel, or prefabricated concrete poles are usually adequate for mounting small turbines for electric generation, since their solidity is low. Bigger turbines, or high solidity rotors for water pumping, may employ steel lattice structures, having either triangular or square plant, or a steel pole with large circular section.

In some cases, it is necessary to stiffen the support tower with steel stays or struts, in order to minimize the oscillations of its top and the fatigue stress deriving from such vibrations. Slender lattice beams can be employed instead of circular section poles when the main issue is keeping the total weight as low as possible.

In industrialized countries, the simplest—and often the cheapest—solution is to build the turbine's support structure with a standardized steel pole or tower, like the ones employed in public lighting and medium/low voltage distribution lines. In rural areas or in developing countries, wooden or concrete poles may be cheaper and easier to procure. The tower's height is a function of the context in which it will be installed. The presence of obstacles, such as big trees or tall buildings, will perturb the airflow, creating vortexes that in general reduce the performances of all kinds of turbines. It is then very important to pay much attention when choosing the site to erect the support structure, in order to keep the turbine at a minimum distance from the surrounding obstacles, and moreover, so that the rotor's lowest point is placed higher than the highest obstacle. As a guideline, Figure 7.1

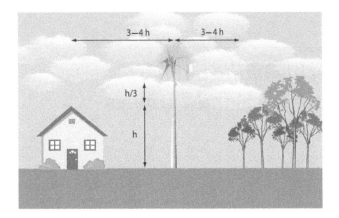

FIGURE 7.1
Minimum relative distances as a function of the obstacles' heights. Drawing by arch. Giovanna Barbaro, based on a sketch by the author.

illustrates the minimum distances from the most common obstacles, resulting from the Author's experience.

The design wind speed is usually defined on available anemometric data. Were such data are not available the International Electrotechnical Commission (IEC) sets international standards for the wind speeds that turbines must withstand. Adopting one of the IEC wind classes shown in Table 7.1 can be then a good starting point.

7.2 Calculation Procedure of a Wind Turbine's Support Structure

7.2.1 Determination of the Loads

Dimensioning a structure requires you to define first all possible loads acting on it, and their combinations. Knowing all stresses on each section of the

TABLE 7.1

Wind Classes according to the International Standard IEC 61400

IEC Wind Classes (m/s)	Class I (High Wind)	Class II (Medium Wind)	Class III (Low Wind)	Class IV (Very Low Wind)
Max. yearly wind speed	50	42.5	37.5	30
Max. annual average wind speed	10	8.5	7.5	6
50-year return gust	70	59.5	52.5	42
1-year return gust	52.5	44.6	39.4	31.5

structure, an adequate safety coefficient can be applied. The loads acting on the structure are only two: the axial thrust on the rotor and the thrust of the wind on the pole. The first is assumed as acting on the hub of the turbine and the second on the barycenter of the support pole. For the determination of the maximum load in the worst conditions, it is necessary to check two cases:

a. The rotor is blocked because of excessive wind speed, and,
b. The turbine is running at its nominal power, at the maximum allowed wind speed.

7.2.1.1 Maximum Load on the Hub under Limited Operational Conditions

In the case of the aerodynamic action turbines, like the *Darrieus*, and horizontal axis ones, operating at the maximum power, the formula explained in Section 2.3.3.5 is fully valid:

$$C_P = C_F$$

From the same, it is possible to deduce C_F for each operation point of the turbine. Another option is to employ the definition of power:

$$P = F \cdot V$$

In the case of the *Savonius* turbine, instead, the following formula must be employed:

$$F = 1.085 \cdot S \cdot V^2$$

Where S is the area of the rotor, and the coefficient 1.085 derives from the expression of the aerodynamic drag forces acting on a concave semi-cylinder ($C = 2.3$) and a convex one ($C = 1.2$), assuming the rotor is blocked (the worst condition for the *Savonius* rotor).

7.2.1.2 Maximum Load on the Hub with Blocked Rotor

It is necessary to check the condition of maximum wind speed and blocked rotor, assuming the blades of the latter as flat plates, with the following formula:

$$F_a = \frac{\rho}{2} \cdot S \cdot 1.5 \cdot V^2$$

Where: S is the area of the blades exposed to the action of the wind (typically 5% of the rotor's swept area in the case of fast horizontal axis turbines, 90% for slow horizontal axis wind turbines, and 14% in the case of *Darrieus* turbines).

7.2.1.3 Load Acting on the Support Structure

The maximum loads on the support structure are calculated for the worst condition, i.e., at the maximum historical wind speed of the site (or the maximum design wind speed according to the local construction code), using the following formula (Figure 7.2):

$$F_{pole} = \frac{\rho}{2} \cdot \left[\frac{h \times (D+d)}{2} \right] \cdot 1.2 \cdot V^2$$

Where:

F_{pole} = force acting on the geometric center of the structure, (N)
ρ = density of air = 1.23–1.25 kg/m³
h = height of the support structure (m)
D = diameter at the base of the support structure (m)
d = diameter at the top of the support structure (m)
V = speed of the wind (m/s)
1.2 = drag coefficient of a cylinder

Combining all the constants, we get:

$$F_{pole} = 0.372 \times h \cdot (D+d) \times V^2$$

7.2.2 Choice of the Pole

7.2.2.1 Standard Steel Poles

Standard steel poles, either conic or tapered ones, are widely employed for public lighting installations and medium/low voltage distribution lines, so they are a commodity easily available in most countries. Their advantages on lattice structures, wooden and concrete poles are multiple:

- Being tubular, it is easy to install the electric cables inside them, and furthermore, they are usually delivered with built-in connection board and cover;
- Their sizes are standardized and hence the selection from a catalog is easy;
- They are aesthetically more agreeable than other solutions.

Steel poles can be either conic or tapered, in general with circular or, more rarely, octagonal section. Models higher than 15 m are usually assembled on site, by joining to each other several conical segments. In the first case, the most frequent when installing small turbines for electric production, it is necessary to calculate the stress at the base section and check that it is smaller than the admissible tension of steel. The stress results from the following formula, valid for a tubular element:

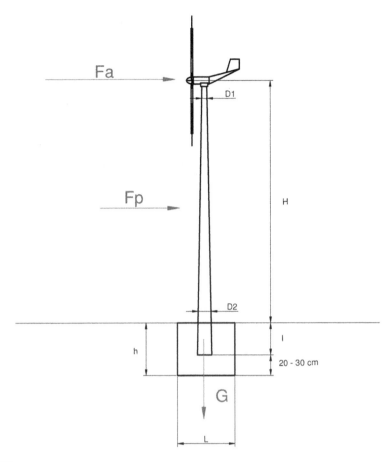

FIGURE 7.2
Sketch of the loads acting on the structure and relative dimensions. See Table 7.1 for the references.

$$\sigma_{max} = \frac{64 \cdot M \cdot D_{ext}}{2 \cdot \pi \cdot (D_{ext}^4 - D_{int}^4)}$$

By combining all of the constants in one, said formula becomes:

$$\sigma_{max} = \frac{10.2 \cdot M \cdot D_{ext}}{(D_{ext}^4 - D_{int}^4)}$$

Where:
D_{ext} = external diameter (m)
D_{int} = internal diameter (m)
σ_{max} = maximum stress on the base joint's section (N/m²)
M_i = bending moment at the base joint's section (Nm)

The bending moment at the base joint's section is calculated with the following formula:

$$M_i = F_a \cdot h + F_{pole} \cdot \frac{h}{2} = h \cdot \left(F_a + \frac{F_{pole}}{2} \right)$$

In order to calculate the value of σ_{max}, it is necessary to retrieve the internal and external diameters at the base joint from the manufacturer's data sheet. Table 7.2 shows an example of such data, corresponding to steel poles for lighting according to Italian norms. Please note that in this example, some heights are available in different models, having different thickness. In order to optimize the cost, it is necessary to choose the model with the minimum section that complies with the condition $\sigma_{max} < \sigma_{admissible}$. The value $\sigma_{admissible}$ is usually specified in the manufacturer's technical sheet and applicable norms. For instance, European structural steels can have $\sigma_{admissible}$ ranging from 120 MPa (low grade) to 240 MPa (high grade). The data presented in Tables 7.2 and 7.3 correspond to poles manufactured with steel having

TABLE 7.2

Selection of Tapered Galvanized Steel Poles, Standard Models (Italian UNE Norm)

H	D_2	D_1	S	L_t	I	G
3.5	89	60	3.0	4,000	500	25
4	89	60	3.0	4,500	500	27
4.5	89	60	3.0	5,000	500	29
5.5	102	60	3.0	6,000	500	39
6	114	70	3.0	6,800	800	52
6	127	70	3.0	6,800	800	57
6	139	70	3.0	6,800	800	62
7	127	70	3.0	7,800	800	67
7	139	70	3.0	7,800	800	69
8	127	70	3.0	8,800	800	72
8	139	70	3.0	8,800	800	77
9	139	70	3.0	9,800	800	84
10	139	70	3.0	10,800	800	92
8	127	70	4.0	8,800	800	80
9	139	70	4.0	9,800	800	94
10	139	70	4.0	10,800	800	102
10	152	70	4.0	10,800	800	107
11	152	70	4.0	11,800	800	132

H = height out of earth (m)
D_2 = diameter at the base joint (mm)
D_1 = diameter at the top (mm)
S = thickness (mm)
L_t = total length of the pole (mm)
I = burial depth of the pole's base (mm)
G = weight (kg)

TABLE 7.3

Selection of Tapered Galvanized Steel Poles. Heavy Duty Models (Italian UNE Norm)

H	D_2	D_1	S	L	I	G
6	152	89	4.0	6,800	800	84
7	152	89	4.0	7,800	800	94
8	152	89	4.0	8,800	800	102
9	152	89	4.0	9,800	800	109
6	168	89	4.0	6,800	800	90
7	168	89	4.0	7,800	800	102
8	168	89	4.0	8,800	800	109
9	168	89	4.0	9,800	800	129
10	168	89	4.0	10,800	800	136
11	168	89	4.0	11,800	800	152
12	168	89	4.0	12,800	800	165
13.2	168	89	4.0	14,000	800	199
6	193	114	4.0	6,800	800	114
7	193	114	4.0	7,800	800	126
8	193	114	4.0	8,800	800	138
9	193	114	4.0	9,800	800	150
10	193	114	4.0	10,800	800	159
11	193	114	4.0	11,800	800	176
12	193	114	4.0	12,800	800	190
13.2	193	114	4.0	14,000	800	230
8	219	114	5.0	8,800	800	176
9	219	114	5.0	9,800	800	183
10	219	114	5.0	10,800	800	194
11	219	114	5.0	11,800	800	214
12	219	114	5.0	12,800	800	231
13.2	219	114	5.0	14,000	800	293
14.2	219	114	5.0	15,000	800	310

$\sigma_{admissible} = 160\,N/mm^2$ (160 MPa). Please note that some norms specify the steel grade according to its elastic limit, σ_{yield}. For practical purposes, assume the following relationship:

$$\sigma_{admissible} = \frac{\sigma_{yield}}{1.5}$$

7.2.2.2 Wooden Poles

Wooden poles are readily available at low cost in most countries. Nevertheless, they present several drawbacks for the installation of wind turbines:

- The maximum height is usually 10–12 m;
- They need to be treated with creosote or other biocide substances in order to avoid rotting;

- A plastic or metallic tube must be fixed to the pole, in order to pass the cables through; and
- A water-tight box containing the connectors must be installed at the base.

The calculation follows exactly the same procedure as the standard steel poles, but the formula of the stress at the base joint is a different one, because the section is solid. The maximum stress in a solid circular section is given by the following formula:

$$\sigma_{max} = \frac{64 \cdot M_i}{2\pi \cdot D^3}$$

By combining all of the constants in one, said formula becomes:

$$\sigma_{max} = \frac{10.2 \cdot M_i}{D^3}$$

Where:
D = diameter at the base joint (m)
σ_{max} = maximum stress on the base section (Pa = N/m²)
M_i = bending moment at the base section (Nm)

The maximum admissible bending stress at the base joint depends on the species of wood, its moisture, the age of the tree, and eventual treatments like creosote impregnation. Table 7.4 provides practical values of admissible bending stress for poles of several wood species, with a safety coefficient equal to 5 (i.e., 20% of the ultimate strength of the pole is subject to cantilever bending).

TABLE 7.4

Admissible Bending Stress for Timber Poles of Several Species (20% of the Ultimate Bending Stresses Retrieved from Several Sources)

Wood Species	Admissible Bending Stress (MPa)	
	Wet	Dry
Yellow birch (Betula alleghaniensis)	9.77	11.5
Douglas fir (Pseudotsuga menziesii)	9.77	11.5
Larch (Larix sp.)	9.77	11.5
Red maple (Acer rubrum)	8.76	10.3
Black oak (Quercus velutina)	9.44	11.1
White pine (Pinus strobus)	8.43	9.92
Redwood (Sequoia sempervirens)	9.1	10.7
Eucalypt (Eucalyptus grandis)	11	12.5
Monterrey pine (Pinus radiata)	10.6	12.6
Coconut palm (Cocos nucifera)	13	16
Chestnut (Castanea sativa)	7	8.3

7.2.2.3 Prefabricated Concrete Poles

There are two types of concrete poles: with H section and with annular section. The second are the most convenient for our purpose, because they usually have built-in earth connectors and, being hollow, facilitate the installation of the cables. Manufacturers of such poles usually specify in their catalogs the collapse load, defined as a horizontal load applied at the top of the pole, that will break its base because of the bending stress. The calculation process is straightforward:

  a. Calculate the maximum thrust at the turbine's hub.
  b. Calculate the wind's load on the pole for the corresponding wind speed, assumed as applied at the pole's center.
  c. Sum of the thrust at the hub plus half of the wind load on the pole.
  d. Look in the table for a model whose collapse load is at least 1.5 times or twice the maximum load obtained in c).

Table 7.5 is an example taken from the catalog of an Ecuadorean manufacturer. Observe that, for similar heights, concrete poles are nearly 10 times heavier than galvanized steel poles.

7.2.3 Sizing of the Foundations

In order to simplify the calculation, the foundation block of a pole is considered as a perfect joint, i.e., the pole transmits all the stresses at its base to a nearly cubic or prismatic concrete block, and the whole system is assumed as perfectly rigid. The resisting moment is given by the weight of the structure plus the elastic reaction of the soil. It can be calculated with the following empirical formula:

$$M_f = 0.85 \cdot P \cdot \frac{L}{2} + k \cdot L \cdot h^3$$

Where:
  M_f = resisting moment of the foundations (Nm)
  P = total weight (foundations block + pole + turbine) (N)
  L = side of the foundation block (assumed prismatic) (m)
  h = depth of the foundation block (m)
  k = 11,000 (empirical coefficient, valid for most soils, even swampy ones)

The depth of the foundations block, h, is calculated as the grounding depth, I, recommended in the catalog of the pole (usually 10% of its total length), plus 0.20–0.30 m.

All factors of the formula are then known, except for L. The minimum value of L that fulfills the condition $M_f > M_i$ under maximum load, is then the

TABLE 7.5

Technical Data of Prefabricated Concrete Poles with Annular Section and Conical Shape

Height (m)	Top Diameter (mm)	Base Diameter (mm)	Collapse Load (N)	Approx. Weight (kg)
14	110	300	4,900	1,054
15	110	310	4,900	1,168
16	110	320	4,900	1,286
17	110	330	4,900	1,407
14	110	300	5,880	1,092
15	110	310	5,880	1,211
16	110	320	5,880	1,335
17	110	330	5,880	1,462
18	110	340	5,880	1,594
19	110	350	5,880	1,730
14	110	300	7,840	1,178
15	110	310	7,840	1,303
16	110	320	7,840	1,433
17	110	330	7,840	1,567
18	110	340	7,840	1,706
19	110	350	7,840	1,848
20	110	360	7,840	1,995
21	110	370	7,840	2,147
22	110	380	7,840	2,305
14	110	300	9,800	1,309
15	110	310	9,800	1,446
16	110	320	9,800	1,588
17	110	330	9,800	1,733
18	110	340	9,800	1,884
19	110	350	9,800	2,038
20	110	360	9,800	2,197
21	110	370	9,800	2,360
22	110	380	9,800	2,530

optimum. Quite often, the optimum side L will be of the same order of magnitude of the pole's diameter at the base. In such cases, just assume $L > 3D$.

7.2.4 Guyed Masts and Towers

7.2.4.1 Generalities

When employing very tall and slender conic or tapered poles, or constant section lattice poles (masts), it is convenient to restrain the structure with guys (more rarely with struts). Such solution is usually employed for erecting anemometric towers, necessary for gathering data prior to the engineering of wind turbine farms. Guyed masts are employed seldom for the installation of wind turbines, limited to very small ones. Guys, also called stays, restrain the mast from falling or bending from at least three sides (Figure 7.3). Guys must be attach to the mast at least at the top, but more commonly several sets of guys, called orders, are attached at different heights (Figure 7.4) to prevent buckling and vibrations. Masts restrained from three or four sides (called Y and X layouts) are intrinsically unsafe in case an anchor breaks; five or more stays will prevent the tower from falling in case of anchor failure. Guys are usually anchored to the ground by means of concrete blocks, more rarely by means of steel stakes.

Masts restrained by two or more orders of stays are hyperstatic structures, so their engineering requires complex methods, the explanation of which is beyond the scope of this book. If a very tall and slender guyed mast is the only possible solution for installing a wind turbine, then engaging a competent civil engineer for designing it becomes mandatory.

7.2.4.2 A Simplified Calculation Method: Range of Validity and Description

The method and "rules of thumb" presented in this section can be employed if, for some reason, a small turbine must be installed at such height that building a lattice tower or employing a non-standard pole results prohibitive, or if only plumbing steel piping is available in the area. Though standard galvanized steel pipes can be as big as 24″ (610 mm) in diameter, finding pipes bigger than 6″ (168 mm) is difficult in some places, and in any case, the maximum allowable stress for mild steel is 103 MPa. Even if the pipe may resist the static loads induced by the wind, a slender section will cause the support structure to bend and vibrate with high amplitude, increasing the risk of collapse by fatigue. The calculation method explained in this section can safely solve such problem, provided it is limited to the following restrictions:

a. The turbine is installed in an open flat site, away from other constructions, and people do not usually stay in the surroundings.

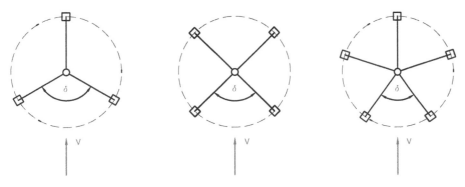

FIGURE 7.3
Guyed masts layout.

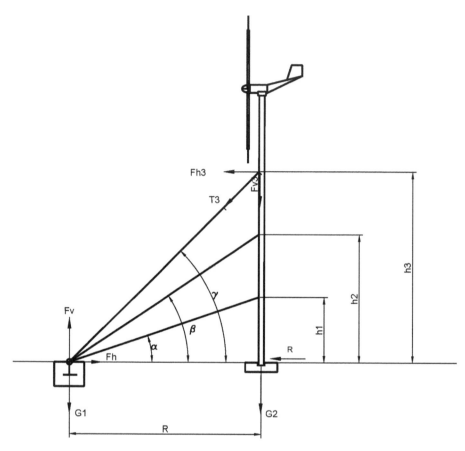

FIGURE 7.4
Vertical distribution in three orders of guys, and the decomposition of the resultant forces in vertical and horizontal loads.

b. Agricultural machines must not operate in the area, in order to avoid accidental damage to some anchoring. Animals pasturing are not an issue.

c. The layout is at least of the X-type, with at least three orders of guys, with the highest one placed at 80% of the total height, or above.

d. The maximum angle between the highest guy and the horizontal line is 60°.

e. The base is a concrete slab with a shallow hole in which the mast fits loosely (Figure 7.4).

f. The initial tension on the stays (no wind load) will increase from the top order to the bottom according to the following "rule of thumb": 5% of the cable's admissible traction for the highest order guy and

20% for the lowest, the remaining ones having equally distributed tensions. This is to avoid resonance with the von Karman vortex trail (see Section 7.2.4.3).

The wind loads acting on the support structure are calculated with the following formulas:

$$F_{pole} = 0.372 \cdot h \cdot (D+d) \cdot V^2 \quad \text{for cylindrical or conical poles}$$

$$F_{pole} = 0.0744 \cdot h \cdot (D+d) \cdot V^2 \quad \text{for lattice poles}$$

The calculation algorithm involves checking the resultant forces on the highest order stay, as if the others did not exist (assuming hence that the structure is isostatic). In the worst wind condition (wind blowing at the highest speed from the direction shown in Figure 7.3), the tension resulting from the wind load, plus the initial tension, T_i, must be smaller or equal to the maximum admissible tension of the cable, T_{max}. As in any problem of statics, the sum of the vertical and horizontal forces and of the moments referred to a point must be zero. With such conditions, it is easy to derive the following formulas (please refer to Figures 7.2–7.4 for the meaning of the symbols).

a. With no wind, the initial (installation) tension of the highest order guy is 5% of T_{max}, defined as:

$$T_{max} = \sigma_{admissible} \cdot S_{cable}$$

where:
  S_{cable} = cross section of the cable
  $\sigma_{admissible}$ = the admissible stress of the cable's material, usually 120–160 MPa for steel cables

For instance, if the structure has three orders of guys like, the one shown in Figure 7.4, the initial horizontal and vertical components are:

$$F_{hi3} = 0.05 T_{max} \cos(\gamma); F_{vi3} = 0.05 T_{max} \sin(\gamma)$$

b. With wind, the total tension of the guys on the side upwind will be the sum of the initial tension plus the components resulting from the wind, while the guys downwind will only keep their initial tension. The sum of the horizontal components must be zero, hence:

$$\frac{(F_a + F_p)}{2\cos\left(\dfrac{\delta}{2}\right)} - F_{h3} - R = 0$$

Where R is the horizontal reaction at the base, assumed as an isostatic hinge.

The sum of the moments with respect to the base must be null too:

$$\frac{F_a}{2\cos\left(\dfrac{\delta}{2}\right)} \cdot h + \frac{F_p}{2\cos\left(\dfrac{\delta}{2}\right)} \cdot \frac{h}{2} - F_{h3}h_3 = 0$$

From this latter, we calculate F_{h3}, so the tension on the cable caused by the wind is:

$$T_3 = \frac{F_{h3}}{\cos\gamma} \rightarrow F_{v3} = T_3 \sin\gamma$$

The total tension must be smaller than the admissible tension:

$$T_{3\text{total}} = T_3 + T_i < T_{\max}$$

And finally, the vertical load on the anchor block will be:

$$F_v = T_{3\text{total}} \sin\gamma$$

In order to avoid that a strong wind tears the anchor block up, the weight of the latter, G, must be at least n times F_v, where n is the number of orders of guys ($n = 3$ in this example).

7.2.4.3 Wind-Induced Vibrations and Fatigue Stress

From the aerodynamic point of view, circular sections are blunt bodies, i.e., wind will produce wake vortexes when flowing around a cylinder placed perpendicular to it. Such vortexes detach alternately from the cylinder's surface, with a frequency that depends on the diameter and wind speed. Such wave of vortexes is the equivalent of a pressure wave, which we can perceive from cables and antennas when the wind is very strong. The first to study the vibration of telegraphic cables in wind was the Czech physicist Vincent Strouhal, by the end of the nineteenth century. He derived an empirical formula linking wind speed and diameter of a cable or wire suspended perpendicular to the flow's direction:

$$\frac{f \cdot d}{V} = St$$

Where:
  f = Frequency of the sound produced by the wind (Hz)
  d = diameter of the wire (m)

V = speed of the fluid perpendicularly to the wire (m/s)
St = Strouhal's number ≈ 0.19

Later on, Otto von Karman discovered that the Strouhal's number is not constant, and derived a complete expression of the frequency of the vortexes trail, valid for any fluid, known as von Karman's formula:

$$\frac{f \cdot d}{V} = 0.198\left(1 - \frac{19.7}{Re}\right)$$

Where:
f = Frequency of detachment of the vortexes (Hz)
d = diameter of the cylinder (m)
V = speed of the fluid perpendicularly to the cylinder (m/s)
Re = Reynolds number

We recall that, by definition:

$$Re = \frac{V \cdot d}{v}$$

and:
v = kinematic viscosity = $14.9 \times 10^{-6}\,m^2/s$ (air at 20°C)

von Karman's formula is valid for $Re > 250$, which is usually the case in most practical conditions.

von Karman's vortexes become audible pressure waves when the diameter of the cylinder exposed to the wind is smaller than 50 mm, which is usually the case with steel guys employed to build support masts. The thinner the cable, the higher the pitch, and hence the more annoying the wind's whistle for humans and animals.

On the other side, a cable or wire suspended with a given tension has a natural vibration frequency, given by the following formula:

$$f_n = \frac{n}{2L}\sqrt{\frac{T}{\rho}}$$

Where:
n = 1, 2, 3... are the vibrational harmonics
L = length of the string (m)
T = tension on the string (N)
ρ = linear mass of the string (kg/m)

If the von Karman vortex frequency is equal to the fundamental or to one of the harmonics of the tensioned cable, the latter will enter into resonance.

The amplitude of the vibration will (theoretically) grow limitless, bringing the cable to break because of excessive strain. In order to avoid resonance phenomena, the guy must be kept loose, as explained in the former section. In that way, the natural vibration frequency of the guy will be very different from the von Karman frequency, and no resonance will arise. Since the times of Strouhal's research, it is usual practice not to tension telephone wires or electric cables more than 25% of their elastic limit. The same criterion is valid for stays supporting masts for small wind turbines.

If the wind turbine is mounted on a pole with circular section, the same will create von Karman vortexes too, but the frequency will be very low (inaudible) if the pole's diameter is bigger than 100 mm. The mass of the turbine mounted on the top of the pole will dump the vibrations, because it represents a discontinuity in the linear mass distribution of the pole. The natural oscillation frequency of a mast carrying a wind turbine on its top is usually much lower than the frequency of the von Karman vortexes, so resonance is impossible.

7.2.5 Foldable or Hinged Poles

Figures 7.5 and 7.6 illustrate two interesting solutions proposed by the Italian manufacturer *Interwind*®. Both systems are very easy to erect and collapse back in case of need (e.g., for maintenance). The bases are hinged and a crank system facilitates the safe operation.

7.3 Practical Exercises

7.3.1 Design of the Support Pole and Foundation Block of a Wind Turbine

A 3-bladed turbine, $D = 2$ m, yields its maximum rated power, 1,750 W, at the speed of 14 m/s. At higher winds, a safety mechanism halves the power and at 29 m/s an emergency break blocks the rotor. The survival speed of the blocked rotor is stated as 30 m/s. Check the axial load in each of the cases. The turbine weights 50 kg (500 N).

The pole (see Table 7.3) is 10 m high, with diameter at the top $d_1 = 89$ mm and diameter at the base $d_2 = 168$ mm. Its thickness is 4 mm. Its burial depth is 800 mm. Its weight is 186 kg (1,860 N) and the manufacturer states $\sigma_{adm} = 160$ MPa. Check if the pole can withstand the maximum survival wind speed and design an adequate foundation.

Solution
Under nominal operative conditions, we get:

FIGURE 7.5
Collapsible mast of the Italian manufacturer *Interwind®*, suitable for mounting on flat roofs. Photo from the catalogue, www.interwind.it.

$$F_a = \frac{1,750(\text{W})}{14(\text{m/s})} = 125(\text{N})$$

Under the limit operation condition, the force will be:

$$F_a = \frac{875(\text{W})}{29(\text{m/s})} = 30(\text{N})$$

Under extreme conditions (blocked rotor), the force will be:

$$F_a = 5\% \left[\frac{1.24}{2} \cdot 1.5 \cdot (\pi \cdot 1^2) \cdot 30^2 \right] = 131(\text{N})$$

In this case, the biggest force is the one acting on the blocked rotor with the strongest design wind. In real life, it is unlikely that the wind ever reaches 30 m/s in most sites, so it may happen that the maximum design load is

FIGURE 7.6
System of collapsible guyed structure of the Italian manufacturer *Interwind®*, suitable for the installation on the ground. Photo from the catalogue, www.interwind.it.

reached under nominal operation or in intermediate conditions. For said reason it is necessary to check always all cases.

Now we can calculate the thrust on the pole under the worst condition:

$$F_{pole} = 0.372 \cdot h \cdot (D+d) \cdot V^2$$

$$F_{pole} = 0.372 \cdot 10 \cdot (0.089 + 0.168) \cdot 30^2 = 860(\text{N})$$

We assume that the thrust on the pole is applied on its geometric center, hence at half of its height. Figure 7.2 shows a sketch of the loads acting on the structure.

The moment at the base is hence:

$$M_i = F_a \cdot h + F_{pole} \cdot \frac{h}{2} = h \cdot \left(F_a + \frac{F_{pole}}{2} \right)$$

$$M_i = 10 \cdot \left(131 + \frac{860}{2} \right) = 5,610 \, \text{Nm}$$

Now we can check if the section and material composing the pole can resist such moment:

$$\sigma_{max} = \frac{10.2 \cdot M \cdot D_{ext}}{(D_{ext}^4 - D_{int}^4)}$$

$$\sigma_{max} = \frac{10.2 \cdot 5,610 \cdot 0.168}{(0.168^4 - 0.160^4)} = 68,066,358 \, Pa = 68.1 < 160 \, MPa$$

Hence the pole will be strong enough under the worst wind condition.

Now we can calculate the size of the concrete foundation. We recall from Section 7.2.3 that:

$$M_f = 0.85 \cdot P \cdot \frac{L}{2} + k \cdot L \cdot h^3$$

Where:

M_f = resisting moment of the foundations (Nm)
P = total weight (foundations block + pole + turbine) (N)
L = side of the foundation block (assumed prismatic) (m)
h = depth of the foundation block (m) = burial depth + 0.2 m
k = 11,000 (empirical coefficient, valid for most soils, even swampy ones)

Since the weight of the foundation is not known in advance, we can proceed by trial and error. Assume $L \approx 3 \times D$ as start point. The density of concrete is 2,000 kg/m³, hence the weight of a prism $0.5 \times 0.5 \times 1$ m³ will be 500 kg = 5,000 N.

$$M_f = 0.85 \cdot (5,000 + 500 + 1,860) \cdot \frac{0.5}{2} + 11,000 \cdot 0.5 \cdot 1^3 = 7,032 \, Nm$$

The foundation's resisting moment is bigger than the capsizing moment caused by the wind on the structure, so the foundation is adequate.

7.3.2 Wooden and Concrete Poles

Using the spreadsheet *structures.xlsx* provided in the download zone, solve the same problem using a 14 m-high concrete pole and a 10 m-high wooden pole.

7.3.3 Guyed Mast

The same turbine of the former example will be mounted on top of a pole, $h = 20$ m. The place is a remote rural area. Standard 4″ (100 mm), 3 mm thick steel pipe is available from the only hardware shop in the region. Since such tube is too slender, the structure will be guyed with three orders of steel cables, disposed in X plant, $17 \times 17\,m^2$ (see Figures 7.3 and 7.4).

Calculate the section of the steel guys and the size of the anchor blocks. Use the spreadsheet *structures.xlsx*.

Solution

The thrust on the wind turbine has already been calculated in the former example, the calculation of the thrust on the steel tube is straightforward, so:

$$F_a = 131N$$

$$F_p = 1,339N$$

Suppose that the cables have 5 mm diameter and are made of mild steel.

$$T_{max} = 160N/mm^2 \cdot 19.6mm^2 = 3,141N \rightarrow T_i = 0.05 \cdot T_{max} = 157N$$

Suppose that the guys are attached to the pole at 18.5, 12.5, and 6.5 m, respectively. The choice of fractions of the total pole's length that are not integers is to avoid restraining the pole at it vibrational nodes. The calculation of the distance to the anchoring blocks and the resulting angles is straightforward and left to the reader as exercise. The base is wide enough, since the angle of the highest guy is smaller than 60°. Applying the formulas explained in Section 7.2.4, the resulting horizontal and vertical force components on the highest guy are:

$$F_{h3} = 612N$$

$$F_{v3} = 943N$$

hence $T_{total} = 1,282N < T_{max}$. The anchor blocks must be cubes having 0.7 m side (6,860 N of weight). The concrete cubes can be directly casted in holes dug directly in the ground.

7.3.4 von Karman Vortexes and Resonance Phenomena

Using the spreadsheet *resonance.xlsx* provided in the download section, check if any of the three orders of guys in the former example has risk of entering in resonance with the von Karman vortexes.

Solution

The spreadsheet is self-commenting. It is necessary to calculate the length of the different orders of stays and enter their corresponding tension. Resonance phenomena are a problem at low wind speed, because such condition represents most of the operation time and consequent accumulation of fatigue. Unless you prefer calculating the real tension at each wind speed with the spreadsheet *structures.xlsx*; assume 1.20 times the initial tension T_i calculated in the former example and check the wind interval up to 10 m/s. Please note that the figure applicable to the vibration formula is the elastic limit σ_{yield}, and not the static limit $\sigma_{admissible}$ (which has a safety coefficient). Also note that the relevant harmonics that may evolve sensible resonance are the fundamental and the three first order harmonics. Higher order harmonics have decreasing amplitudes, so even if some wind speed can induce von Karman vibrations of the same frequency of, for instance, the fifth or tenth harmonic, such situation is not dangerous. Observe that, for instance, a cable having 5 mm diameter, 25 m long, tensed at 600 N, has 1.24 Hz of natural resonant frequency. The von Karman vortex trail would produce such frequency when the wind is nearly 0.09 m/s, but the Reynolds number is too low for the formula to be valid, so the intensity of the vortexes will be very weak, or null. Relevant vortexes require at least 1 m/s to exist, but their induced frequency is much higher than the fundamental and the first three harmonics of the tense cable, so we can conclude that resonance between the vortexes and the stays of our example is virtually impossible. On the contrary, winds faster than 1 m/s will produce audible noise. Nevertheless, the pitch is low; hence, it is unlikely to become bothering for humans or animals.

Bibliography

Arancon R.N. Jr., *Asia-Pacific Forestry Towards 2010*, Food and Agriculture Organization of the United Nations Working Paper No: APFSOS/WP/23, Forestry Policy and Planning Division, Rome, October 1997.

Elecdor, *Catalogue of Steel-Reinforced Concrete Poles for Electrical Lines According to Ecuadorian Norms*, www.elecdor.ec.

Rosato M., *Diseño de máquinas eólicas de pequeña potencia*, Editorial Progensa, Sevilla, Spain, 1992.

Rosato M., *Progettazione of impianti minieolici*, multimedia course, Acca Software, Montella (AV), Italy, 2010.

Torrán E.A., Sosa Zitto M.A., Cotrina A.D., and Piter J.C., Bending strenght and stiffness of poles of Argentinean *Eucalyptus grandis*, Maderas-Ciencia y Tecnologia 11(1) Concepción, Argentina, 2009, doi:10.4067/S0718-221X2009000100006.

Wolfe R.W. and Kluge R.O., *Designated Fiber Stress for Wood Poles*, United States Department of Agriculture, Forest Service, Forest Products Laboratory, General Technical Report FPL–GTR–158, June 2005.

8

Probability Distribution of the Wind Speed and Preliminary Design of Wind Power Installations

8.1 Generalities

One of the most controversial and delicate aspects in the design of wind power systems, is the preliminary estimation of the potential energy generation in a given place, and its distribution along time. The argument is vast enough to fill a specialized book, and many peer-reviewed papers have been published already, so in this volume we will only study the basic notions. On the other hand, small wind turbines are usually employed for non-critical applications, in environments where the return of the investment is often a second-order argument compared to other requirements, like self-sufficiency, easiness of reparation, and reliability of operation. There are specialized companies that provide databases of meteorological data, with very high geographical resolution (up to $100 \cdot 100\,\mathrm{m}^2$ grids). Such data are called "typical meteorological years" (TMY) of wind speed at a given reference height and represent the most probable wind speed distribution along a whole year. A TMY is the average of measurements taken along 10 or more years, at a specified height above ground. Such a large amount of samples eliminates seasonal and multi-annual cyclic variations (e.g., the five-year climatic cycles known as *El Niño* and *La Niña*). Such data are quite reliable, but often unaffordable for small wind power projects, being the cost directly proportional to the fineness of the geographical resolution. In some cases, purchasing high-quality wind data from specialized companies may cost as much as a small wind turbine.

Most developing countries have very poor wind data. The sole data source available is often meteorology stations and wind measurements at airports, but the accuracy of standard anemometers employed in weather stations is inadequate for assessing the wind energy potential, as will be explained in Section 8.3.3.

Existing wind speed measurements of poor quality may therefore be an insufficient guide to correctly design a wind power installation for a given place.

This is where modern mesoscale wind mapping technology comes into the picture; since it is based on data from earth observation satellites, historical reanalysis data, and global meteorology models. We will not deal with mesoscale modelling theory in this book, but just apply its results in a practical example. A quick guide on mesoscale theory, free to download from the Internet, is included in the Bibliography for the readers willing to learn more about the subject.

From the point of view of the designer and builder of small wind power systems, the alternatives to determine the wind potential in a given place are four:

1. purchasing a TMY with high geographical resolution from a specialized company (quickest and most reliable option, but usually not affordable);

2. obtaining historical series logged by weather stations—but these may be placed at big distances from the intended site of installation, so the historical data are not necessarily applicable to the location and the accuracy of the wind speed measurements may be low;

3. conducting an anemometric campaign, at least one year long, directly on site. This solution will provide true data, but the measurement error margin can be high and analyzing just one year may not be statistically relevant, the cost of the measurement station, anemometric pole, and installation may be of the same order of magnitude of the purchase cost of a small commercial wind turbine;

4. employing probability functions and data on average wind speed, obtained from large territorial scales (for instance, mesoscale maps of average wind speeds and roughness classes). This option will provide "order of magnitude" information. It turns useful to define the nominal wind speed and size of the turbine for a given expected energy productivity.

The last option has the lowest resolution and highest uncertainty margin, but in general it is based on information that can be obtained for free, or at very low cost, from public meteorological agencies, so it will be the usual choice for small projects or feasibility studies. We will see with a practical example that the accuracy is acceptable for small wind power systems.

An aspect that must be taken into account when dealing with wind speed data is how average wind speeds are expressed: scalar average or vector average. The wind is described as having both a direction and a magnitude (speed), and it is therefore a vector quantity. Although the wind is a vector quantity, the wind direction and speed can be treated separately as scalar values. Samples are typically collected at a high frequency and then averaged over a time period of a few minutes to an hour. Depending on the application and the instrumentation employed, the data may be vector averaged, scalar averaged, or averaged using both techniques.

In scalar averaging, the anemometer makes independent measurements of the wind speed and direction at a given sampling rate—e.g., one sample every 30 s. The arithmetic averages of the outputs—wind speed and direction—are calculated over the averaging period.

In vector averaging, either the orthogonal components of the wind are measured directly with a wind instrument or the speed and direction are measured with an anemometer and a wind vane and then they are used to derive the orthogonal components. To obtain the vector-averaged speed and direction, the components are summed and vector averaged at the end of the averaging time.

During periods of moderate to high wind speeds, the difference between vector and scalar averages will be small. As a rule, vector-averaged speeds will be always lower than scalar-averaged values. Larger differences will occur with greater wind direction variance, which typically occurs at lower wind speeds (below about 2 m/s) or in areas of high turbulence (e.g., downwind of mountains or other obstacles). As an extreme example, suppose we had a constant wind from the north at 5 m/s for 5 min followed by a constant wind of 5 m/s from the south for 5 min. If we calculated both the vector and scalar averages for the 10 min period, the vector-averaged speed would be zero, whereas the scalar-averaged speed would be 5 m/s. In the real world, the differences are not so extreme, with scalar averages being between 10% and 20% larger than vector averages. For the productivity analysis of small wind turbines oriented by vanes or conicity, scalar averages are the most suitable data format, because the low inertia of the yaw system will keep the turbine always faced toward the wind. Vector averages are suitable for calculating the productivity of large wind turbine oriented by servomotors, because their high inertia makes impossible to orient the rotor instantly with any small variation of the wind direction.

8.2 Employing "Typical Meteorological Years" from Actual Weather Stations or from Specialized Companies

This technique requires loading the data on a spreadsheet. The ideal dataset should contain at least the hourly scalar average along a "TMY". As a first approach, such data can be obtained from the public meteorological service or from private weather stations, assuming arbitrarily any recent year as being a "TMY." A "true TMY," as the ones employed for the design and assessment of large wind farms, is derived from statistical analysis along longer periods of time, at least ten years, but when there are no other data available, a dataset corresponding to one full year could be enough. We will see in Section 8.3.1 that the instrumental error accumulated in the calculation of the available energy can be very high, so for practical purposes the data corresponding to one full year can be assumed as being a "true TMY." The minimum data that are necessary for calculating the average wind power during each interval of time are the scalar wind speed and, if possible, air temperature, and pressure too. Hourly averages are already a good compromise between good accuracy and size of the dataset. Of course, it is better if half-hour or quarter-hour averages are available, but the size of the file will be bigger. Air temperature, pressure, and eventually also moisture are useful for calculating the density of the air, ρ, at any time interval. If such data are not available, or if you want to keep the analysis as simple as possible, then assume an average air density. In general, anemometers are placed at 10 m height, but this is not an absolute standard. It is necessary to know in advance at what height above ground the anemometer was placed, the latitude, longitude (and altitude if in a mountainous place) of the site. Then, calculate the average output power of the turbine for each time interval with the already known formula:

$$P = \frac{\rho}{2} \cdot C_P \cdot S \cdot V^3$$

where S and C_P are the exposed area and the power coefficient of the turbine (from the manufacturer's catalog of from our own calculations, depending on the case). Adding the hourly powers we can then obtain the total annual energy, or the daily energy distribution, or the monthly average energy distribution, along the TMY. In general, the average hourly power is calculated in two steps: first the hourly wind energy density and then the hourly power output of a concrete wind turbine. The energy density is calculated as:

$$E = \frac{\rho}{2} \cdot V^3$$

and sometimes it is available in some wind maps or other information systems.

An example of analysis with a spreadsheet is provided in Section 8.6.

Since the average power (or energy productivity) results from a calculation on several measured magnitudes, and each magnitude has its own measurement error or uncertainty, it is necessary to estimate the propagation of the total uncertainty, as explained in the next Section 8.3.

8.3 How to Design Your Own Anemometric Campaign

If no wind data are available for the site, a possible alternative is to purchase and install your own weather station. The data that the weather station must be able to collect are the ones already specified in Section 8.2, and at least one year of data logging is necessary, in order to account for eventual seasonal variations. The decision to perform an anemometric campaign must be thoroughly meditated, since a good quality weather station may cost as much as the small wind turbine installation itself, and one year of measurement campaign is one year of energy production lost. In general, the differences of total available energy between different years are small. It is important to remember that weather stations do not measure the wind's energy, but wind speed, and eventually also atmospheric pressure, temperature, and relative moisture. The available energy will then result from a calculation in which such measured magnitudes are the/ factors. Each time that a physical magnitude is derived by calculation from other individually measured magnitudes, the unavoidable measurement uncertainties are amplified. The study of the error propagation is fundamental in Metrology and a series of theorems, known as *Theory of Errors*, allows us to estimate the final uncertainty margin of our calculation's result. There is much specialized literature on this argument, so in the present book we will just recall the theorems and apply them as rules.

8.3.1 Basic Notions of Metrology: Accuracy, Precision, and Repeatability

> I often say that when you can measure what you are speaking about, and express it in numbers, you know something about it; but when you cannot measure it, when you cannot express it in numbers, your knowledge is of a meagre and unsatisfactory kind; it may be the beginning of knowledge, but you have scarcely, in your thoughts, advanced to the stage of science, whatever the matter may be.

These words of Sir William Thomson (better known as Lord Kelvin, and because of the temperature scale nominated in his honor, Figure 8.1), published in 1883, summarize the essence of this chapter. Determining the wind power potential in a given place requires measuring a series of parameters with the scope of estimating the turbine's power output, and its integration along a TMY. But what does "to measure" mean?

> To measure means comparing an unknown physical magnitude to another magnitude of the same type, assumed as a standard.

It is easy to measure a well-defined physical magnitude, as for instance, a length, a weight, or an electric current. It is difficult to measure complex processes such

FIGURE 8.1
Sir William Thomson, baron of Kelvin. Public domain photo from https://commons.wikimedia. org/wiki/File:Portrait_of_William_Thomson,_Baron_Kelvin.jpg.

as the energy produced by the wind, since it is the integral along time of a function depending on the cube of wind speed, on the turbine's C_P (which in turn is variable as a function of λ, which in turn depends on the wind's speed, on the mechanical features of the turbine and its speed regulation system, and on the load on the generator), and on the density of air (which in turn depends on the altitude, pressure, temperature, and absolute moisture).

We can add another truth to Lord Kelvin's words, apparently obvious but not always proving true in the practice: apart from *being able to measure*, it is also necessary to *know how to measure* correctly. Hence, before discussing some practical examples, in the next paragraphs we are going to introduce some theoretical concepts about how to perform reliable measures.

8.3.2 Definitions of Accurateness, Precision, and Repeatability

Figure 8.2 summarizes the definitions of both concepts with the example of a keen sniper who tests four different rifles or, conversely, of four snipers with different degrees of ability testing the same rifle. The snipers can fail in hitting the center of the target. The analogy of the snipers with our context allows us to state that the measured errors can be caused either by an incorrect calibration of the instrument or by the user's inexperience or careless use.

A good instrument must be precise, i.e., it must measure values that are consistent with the time—both in short-term and long-term repeated measures—but it must be accurate too, i.e., it must measure values very close to the true value of the physical magnitude (case A).

An instrument that is precise but not accurate (case B) induces a systematic error, easy to compensate on condition that the amount and sign of the error is known.

An instrument that is accurate, but not precise (case C), induces random errors. Such kind of instruments is acceptable for maintenance scopes, because the average value of the measures will be close to the true value, but will require a relatively high number of measures in order to obtain a representative average value.

An instrument that is neither precise nor accurate (case D) should not be employed, because the measures will be unreliable.

8.3.3 Error Propagation

There are two different situations in which we obtain useful information on the base of measured values. In the first case, we *measure directly* with a *dedicated* instrument the physical magnitude that we are interested in knowing; in the second case, the unknown physical magnitude results from a *calculation from one or more measures*. When measuring directly a physical magnitude, the error (or uncertainty) of the measured value depends directly on the instrument (and/or on the user's skill). When the unknown physical

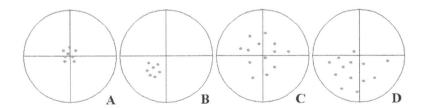

FIGURE 8.2
Graphic analogy of the concepts of accurateness, precision, and repeatability. A: Accurate and precise, B: Precise, but not accurate, C: Imprecise (or disperse), but the average value may be acceptable because the dispersion is not biased in a given direction, D: Imprecise and inaccurate.

magnitude is derived from one or more measures, the total error or uncertainty of the calculated value will be a function (usually a sum or an amplification factor) of the single measures and their respective errors. In order to understand how the errors propagate when a value is derived from several measures, it is possible to employ a series of rules, which are based on theorems. Since the scope of this book is to prioritize the practical experience over the theory, we will not present here the development of the theorems and the error propagation will be estimated by the application of said rules. We are going to start with the definition of the error types and their symbols, presenting a few examples that will make easier to understand how to apply the error propagation rules.

Let us start by defining the symbology and the concepts underlying the practical calculation of the uncertainty in estimating the energy from the wind, when such estimation results from employing meteorological data.

8.3.3.1 Definition N. 1

$E_{(x)}$ = absolute error or uncertainty of the measured magnitude x.

If x_m is the value of a physical magnitude measured in a given way and x_v is the true value (in general unknown), the absolute error $E_{(x)}$ is defined as:

$$E_{(x)} = x_m - x_v$$

$E_{(x)}$ is hence expressed in the same unit of the measured magnitude (m, kg, V, etc.).

Example of absolute error

Consider a barometer having a scale with 1 mbar divisions. The absolute error of the atmospheric pressure measured with such a device is ±1 mbar, because the intermediate values cannot be measured with certainty. It is though possible to observe with naked eye if the needle is between two consecutive divisions, but it is impossible to state if it is in the exact middle.

8.3.3.2 Definition N. 2

$e_{(x)}$ = relative error or uncertainty of the measured magnitude x.

The relative error $e_{(x)}$ is defined as:

$$e_{(x)} = \frac{E_{(x)}}{x_v}$$

Consequently, $e_{(x)}$ is non-dimensional and will be expressed in %.

Example of relative error

Consider the same barometer of the former example. If the measured value is 893 mbar, since the uncertainty of such measure is ±1 mbar, the relative error will be equal to 1/893, i.e., ±0.1%. If instead, we measure 1,060 mbar, the relative error becomes 1/1,060, i.e. ±0.09%.

First practical conclusion

When a physical magnitude is measured with whatever instrument, it is good practice to choose the instrument's scale in order that its maximum range (a.k.a. end of scale or top scale) results as close as possible to the supposed value of the magnitude.

8.3.3.3 Rules of Error Propagation

Now that the concepts of relative and absolute error of a measured value have been defined, the fundamental theorems of the *Theory of Errors* can be stated as the following practical rules:

1. Relative error of the product of two measured magnitudes, $x \cdot y$

$$e_{(x \cdot y)} = e_{(x)} + e_{(y)}$$

2. Relative error of the nth power of the measured magnitude, x. As a consequence of rule n. 1:

$$e_{(x^n)} = n \cdot e_{(x)}$$

3. Relative error of the quotient of two measured magnitudes, x/y

$$e_{(x/y)} = e_{(x)} + e_{(y)}$$

4. Absolute error of the sum or the difference of two measured magnitudes, $x \pm y$

$$E_{(x \pm y)} = E_{(x)} + E_{(y)}$$

Second practical conclusion

From the former rules, we can deduce that the maximum error made when calculating a value with a formula, based on physical variables measured independently, depends on the formula and on the method employed, and not only on the precision and accuracy of the instruments.

8.3.3.4 Conventions for the Correct Expression of Measured Values, or of Values Calculated from Measures, and Their Errors

a. $E_{(x)}$ is always written with a single significant figure.
b. The result of a measure whose $E_{(x)}$ is known, is written with all the significant figures up to the same order of the error, rounded.

8.3.3.5 Estimation of the Errors in the Calculation of the Energy Productivity from Meteorological Data

Since in each single measurement interval the power at the generator's terminals is given by the already known formula:

$$P = \eta \cdot \frac{\rho}{2} \cdot C_P \cdot S \cdot V^3$$

the application of the error propagation rules leads to the following result:

$$e_{(P)} = e_{(\eta)} + e_{(\rho)} + e_{(C_P)} + e_{(S)} + 3 \cdot e_{(V)}$$

The efficiency of an alternator is generally assumed to be greater than 90%, which is more or less true for big alternators ($P > 100\,\text{kW}$), but absolutely false for small ones, especially when they do not operate at full load. The conversion efficiency of the mechanical energy at the shaft of the generator into electrical energy at the terminals of the load depends on the load itself (it is maximum when the power at the terminals is close to the alternator's nominal power), and on the eventual presence of rectifiers, voltage regulator, battery bank, and inverter. Furthermore, it is necessary to consider the overall energy conversion efficiency. If, for instance, the turbine's shaft is coupled to the alternator's shaft by means of a gearbox, or through pulleys, it is necessary to include in the calculation the conversion efficiency of the same. If the alternator's efficiency curves, as well as those of other components of the system, are not known we can estimate conservatively the value of the efficiency's uncertainty as $e_{(\eta)} = 5\%$.

In general, weather stations do not measure directly the density of air, ρ, so it is necessary to calculate it from a complex formula having as variables the atmospheric pressure, the temperature of air, and the absolute moisture. As a first approach, we can neglect the moisture and calculate the air's density with the perfect gases law:

$$p \cdot v = m \cdot \bar{R} \cdot T$$

from which:

$$\frac{p}{\bar{R} \cdot T} = \frac{m}{v} = \rho$$

Since $\bar{R}$ is a universal constant, equal to $287.05\,\text{J/kg·K}$, by definition it has no error, so the error in the measurement of ρ reduces to:

$$e_{(\rho)} = e_{(p)} + e_{(T)}$$

Typically, the error of electronic barometers for weather stations is about 0.1% while the error of the thermometer is usually bigger than 2.5%.

An alternative approach, if only temperature data are available, is to employ "standard air density" values, Table 8.1.

TABLE 8.1

Density of Air at Standard Atmospheric Pressure

Temperature (°C)	Density of Dry Air (kg/m³)	Max. Water Content (kg/m³)
−25	1.423	
−20	1.395	
−15	1.368	
−10	1.342	
−5	1.317	
0	1.292	0.005
5	1.269	0.007
10	1.247	0.009
15	1.225	0.013
20	1.204	0.017
25	1.184	0.023
30	1.165	0.030
35	1.146	0.039
40	1.127	0.051

Source: Danish Wind Industry Association, http://xn--drmstrre-64ad.dk/wp-content/wind/miller/windpower%20web/en/stat/unitsw.htm#roughness, adapted by the Author.

Finally, the simplest approach is to consider constant air density. The wind power industry adopts as standard the density at sea level, at 101.3 kPa, and 15°C. This results in an error of about 10% of the total energy along a year.

The error of C_p is by far more difficult to estimate, because it depends on Re, on the roughness of the blades surface, on the speed control system, and on the rated power and design of the turbine. We can estimate it in the range 5%–10%.

The error of the exposed area is negligible, because it does not vary with time, apart from the thermal dilatations that are absolutely non-influential.

The error of the wind speed's measurement depends not only on the quality but also on the type of anemometer. A well-known manufacturer of weather stations for private use specifies that the standard anemometer included in its "professional" model, equipped with data logger, has a measurement uncertainty equal to ±3 m/s in the measurement range between 2 and 10 m/s (hence an average error in the order of ±30%!), and ±10% for winds between 10 and 56 m/s. The optional anemometer, defined as "high precision" model, has a maximum error equal to ±5% throughout its whole measurement range. In the market of commercial scale wind energy, the standard IEC 61400-12-1:2017, *Wind energy generation systems—Part 12-1: Power performance measurements of electricity producing wind turbines* states that wind speed measures must be performed with a Class 1 cup anemometer, i.e., a cup anemometer having 1% maximum error.

N.B.: Conventionally, the class of an instrument is expressed as the maximum relative error referred to its measurement top of scale. For instance, measurements taken with a class 1 anemometer with speed range 0–20 m/s have a constant absolute error equal to 0.2 m/s throughout the specified range. This means that measuring low wind speeds has bigger relative error than measuring near to the top of scale. The practical conclusion is that, when planning the purchase of an anemometer to conduct an anemometric campaign, we should choose the model with the top scale as close as possible to the maximum operational speed of the turbine.

The said standard provides a series of directions on how to mount the anemometer in order to minimize the aerodynamic interference of the support mast. There are other types of anemometers, based on sound beams, called SODAR, or on laser beams, called LIDAR. Such instruments measure only vector averages, and their accuracy may vary a lot with the site's turbulence and air temperature. The error of cup anemometers varies only with the wind speed and is influenced by the rotor's inertia compared to the wind's acceleration caused by gusts. Such phenomena are easier to correct when performing the calibration.

Based on the considerations discussed above, we can estimate the total error of the wind power productivity calculated with meteorological data as:

$$e_{(P)} = e_{(\eta)} + e_{(p)} + e_{(T)} + e_{(Cp)} + 3 \cdot e_{(V)}$$

$$e_{(P)} = 5\% + 0.1\% + 2.5\% + 10\% + 3.5\% = \pm 33\%$$

Practical Conclusion

Since the wind's energy is a function of the cube of the wind speed, the uncertainty of the calculated energy potential depends strongly on the quality of the anemometric data and on the reliability of the C_p values—either self-calculated or from commercial manufacturers. Hence, conducing an anemometric campaign with a general purpose weather station, or basing our calculations on data from the public weather service, which usually employs anemometers with higher class than the ones prescribed by the IEC 61400-12-1:2017 standard, may lead to high uncertainties in the determination of the available power and total energy productivity of a site. The former considerations justify then that the designer of a small wind power installation adopts an easier and quicker procedure, which will lead to results having a comparable uncertainty level, but at low or no cost.

8.4 Employing Data of Average Speed and Statistical Functions

Based on the considerations in the former section, we will now focus on how to employ statistical functions to estimate the wind's energy potential of a place, based on historical data usually available for free, namely the average wind speed of the site at a given height above ground.

Two are the probability density functions commonly employed for the said scope: Weibull's and Rayleigh's functions. Both generate a distribution of the probability (in practice, the number of hours in a year when the wind blows at each given speed), using as seed datum the average wind speed of the site. The applicability of one or the other function depends on the case: in weak wind zones ($V_{avg} < 5\,\text{m/s}$), or in zones where the wind is unsteady, Weibull's probability distribution function provides more accurate results, while in windy zones ($V_{avg} > 5\,\text{m/s}$) Rayleigh's function is more accurate. If the annual average wind speed is not available from local weather stations, such value can be estimated from wind maps. For instance, in Italy the *Atlante Eolico Italiano (Italian wind atlas)* is accessible to the public from http://atlanteeolico.rse-web.it/viewer.htm, and other countries have similar sites with wind data, either interactive or to download for free. The DTU (Danish Technical University), together with other research partners, has developed the *Global Wind Atlas (GWA)* http://globalwindatlas.com/, a mesoscale model of the whole world. Employing such maps is costless and relatively quick. Considering the large error that wind data from generic weather stations can induce in the calculation of the energy potential of a site, the results provided by mesoscale models are acceptable for the installation of small wind turbines. Of course, if data from a local weather station are available, it is worth to check these too, and then compare results.

8.4.1 Rayleigh's Function of Probability Distribution

The reader can find the whole theory on statistical functions in any University text on the subject. For practical purposes, the following formula allows the analysis of the wind speed distribution along a year with Rayleigh statistical distribution:

$$N_{(v)} = 8,760 \cdot \frac{v}{\sigma} \cdot e^{\left(\frac{v^2}{2 \cdot \sigma^2}\right)}$$

Where:

$N_{(v)}$ = number of hours in a year in which the wind speed is equal to v
8,760 = number of hours in one year
v = wind speed (m/s)
σ = parameter of the function

The parameter of Rayleigh's function is univocally related to the average wind speed of the site, V, by means of the following formula:

$$V = \sigma \cdot \sqrt{\frac{\pi}{2}} \approx 1.253 \cdot \sigma$$

Rayleigh's function is usually employed by the manufacturers of big turbines, since such machines are always installed in windy areas, where this function describes well the frequency distribution of the wind speed. The wind power industry prefers the use of this function for another reason: being a monoparametric function, its results are directly comparable from one site to another.

8.4.1.1 Example of Use of Rayleigh's Distribution

We know that the average speed of the wind in a site is $V = 6\,\text{m/s}$. The place is flat land with steady winds, so Rayleigh probability distribution is assumed as a suitable model.

The first step is to calculate the parameter σ:

$$6 \approx 1.253 \cdot \sigma \rightarrow \sigma = 4.788$$

Once the value of σ is known, tabulating and plotting the Rayleigh distribution with a spreadsheet is straightforward. The curve represents the number of hours per year that the wind will blow at each of the considered speeds. Since wind turbines are stopped for wind speeds smaller than their cut-in or higher than their cut-off speed, it is necessary to calculate and plot the curve just for the operational speed range of the turbine.

Table 8.2 shows the values of v, the energy density E (calculated assuming constant air density $\rho = 1.22\,\text{kg/m}^3$, $S = 1\,\text{m}^2$, and multiplying the resulting power at each wind speed by the corresponding number of hours). Figure 8.3 shows the curves of the probability distributions of both the wind speed and the energy density within the operational interval of wind speeds 4–$14\,\text{m/s}$ or a generic turbine.

8.4.2 Weibull's Probability Function

For practical purposes, we will apply the following formula:

$$N_{(v)} = 8,760 \cdot \left(\frac{k}{c}\right) \cdot \left(\frac{v}{c}\right)^{(k-1)} \cdot e^{-\left(\frac{v}{c}\right)^k}$$

Where:
$N_{(v)}$ = number of hours in a year in which the wind speed is equal to v
$8,760$ = number of hours in one year
v = wind speed (m/s)
k = shape parameter of the curve
c = scale parameter of the curve

The shape parameter, k, and the scale parameter, c, are theoretically related to the average speed, V, by the following relationship:

TABLE 8.2

Rayleigh's Probability Distribution of the Wind Speed, of the Instantaneous Power of the Turbine and of the Energy, in a Place Where $V = 6$ m/s (Input Parameter)

V	6	m/s, Average Speed from Map	
σ	4.788507582	m/s	
N	8,760	h/year	
v (m/s)	h/year	P (W/m²)	E (kWh/year · m²)
0	0	**0.00**	0
1	374	**0.61**	0
2	700	**4.88**	3
3	942	**16.47**	16
4	1,078	39.04	42
5	1,107	76.25	84
6	1,046	131.76	138
7	919	209.23	192
8	757	312.32	236
9	588	444.69	261
10	432	610.00	263
11	300	811.91	244
12	198	1,054.08	209
13	125	1,340.17	167
14	74	1,673.84	125
Annual wind potential ($v > 3$ m/s)			1,962
Average power density (W/m²) = E/8,760 h/year			224

The bold values indicate the interval in which the turbine is not running.

$$V = c \cdot \Gamma\left(1 + \frac{1}{k}\right) \tag{8.1}$$

Where Γ is called "gamma function." This function can be expressed in different equivalent ways. The following formula is one of them, known as Euler's Gamma Function:

$$\Gamma(z) = \frac{1}{z} \prod_{n=1}^{\infty} \frac{\left(1 + \frac{1}{n}\right)^z}{1 + \frac{z}{n}} \tag{8.2}$$

Fortunately, the gamma function is usually a built-in feature of most spreadsheets. In our examples, the gamma function of $(1 + 1/k)$ can be calculated easily using the following syntax: = *gamma(1 + 1/<cell containing the value of k>)*.

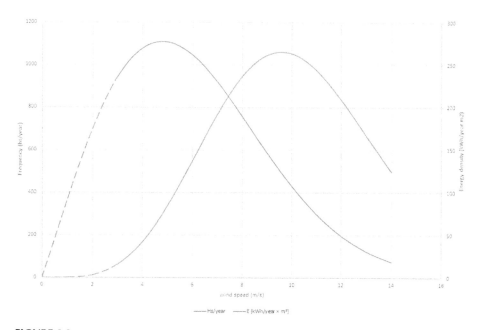

FIGURE 8.3
Rayleigh's probability distribution of the wind speed and of the energy density in a place where $V = 6$ m/s. The dotted portion of the curves indicates the interval in which turbines usually do not produce energy. Observe that the maximum energy density corresponds to $v \approx$ 9.5 m/s and not to the average, V. Consequently, it is convenient to optimize the design of the turbine for such speed in order to get the maximum productivity.

Our main problem when only the average wind speed is known—e.g., from some wind map or other similar source—is that we do not know in advance what the values of k and c will be. The designer of wind power systems is interested more in the available wind energy in a given place, rather than in just the speed distribution of the wind. The mean power density is the total energy of the wind, integrated along the reference period, divided by the same period. The cubic mean speed can be defined as "a constant wind speed that would produce the same energy in a given place in the same period of time." It is possible to demonstrate that the shape and scale parameters of the Weibull distribution of the wind speeds in a site and the corresponding mean power density are related by the following equation:

$$\bar{E} = \frac{\rho}{2} c^3 \Gamma\left(1 + \frac{3}{k}\right) \tag{8.3}$$

Where:
$\bar{E}$ = mean power density (W/m²)
ρ = standard density of air = 1.22 kg/m³

k = shape parameter of the curve
c = scale parameter of the curve
Γ = the gamma function

When the available information, for instance, from the **GWA** (http://www.globalwindatlas.com/) and other similar models, includes both the average wind speed and the mean power density, then the coefficients c and k can be univocally calculated by checking different combinations of c and k until Equations 8.1 and 8.4 are simultaneously satisfied. We will see in detail such procedure in the practical exercise, Section 8.6.

When only the average wind speed V is known, we can just "guess smartly," based on the following rules of thumb:

a. In areas with unsteady winds, for instance, coastal places subject to thermal winds, or high local turbulence (near big cities, mountains, forests), choose $1.3 < k < 1.8$.

b. In windy areas, use values closer to 2, or just employ Rayleigh's function. For instance, winds in the Scandinavian region are well represented by $1.9 < k < 2.2$.

c. In areas with very steady winds, use $2 < k < 3$. For instance, tropical and subtropical areas subject to trade winds are well represented by $k > 2.5$.

N.B.1: defining k with the aforesaid criteria, and then calculating c with Equation 8.1, is a method that should be employed only when no data other than V is available, and/or when there is no time—or economical resources—for performing an anemometric campaign on site. The result obtained with such approach is just an idea of the order of magnitude of the wind power potential in a site.

N.B.2: Trying to extrapolate values of k and c from the literature on big wind turbines and wind farms is an "educated guess" too, because such data usually refer to heights bigger than 100 m, in particular, windy sites. For a given place and local roughness, the value of k grows with the height, reaching its maximum near 100 m, and then decreases with increasing heights. If trying to apply published data from big wind farm projects to a small wind turbine to be placed in their neighborhood, always check if the height considered for the study is higher or lower than 100 m.

Once the values of V, k and c are known, then tabulating and plotting Weibull's function with a spreadsheet is straightforward. The curve represents the number of hours in a year for each wind speed within the operational speed range of the chosen turbine. Table 8.3 shows the values of v, the energy density E (calculated assuming constant air density $\rho = 1.22$ kg/m³, $S = 1$ m² and multiplying the resulting power by the corresponding number of hours), in a place where $V = 2.217$ m/s, $c = 2.432$ and

TABLE 8.3

Weibull's Probability Distribution of the Wind Speed, of the Specific Power and of the Energy Density, in a Place Where the Input Parameters are $V = 2.217\,\text{m/s}$, and $k = 1.4$, c Calculated with the Gamma Function

$k = 1.4$						
$v = 2.217$ m/s						
$c = 2.4324591$						
v (m/s)	v/c	$(v/c)^{(k-1)}$	$\exp(-(v/c)^k)$	N (h/year)	P (W/m²)	E (kWh/year·m²)
0	0	0	1	0	0	0
1	0.4111066	0.700780133	0.749690113	2.649	1	2
2	0.8222132	0.924684929	0.467531683	2.180	5	11
3	1.2333198	1.087502547	0.26152159	1.434	16	24
4	1.6444264	1.220129079	0.134470224	827	39	32
5	2.055533	1.33404286	0.064430436	433	76	33
6	2.4666396	1.434968214	0.029026402	210	132	28
7	2.8777462	1.526233553	0.012374556	95	209	20
8	3.2888528	1.609969972	0.005016844	41	312	13
9	3.6999594	1.687636015	0.001941713	17	445	7
10	4.111066	1.760280107	0.000719735	6	610	4
11	4.5221726	1.8286848	0.00025618	2	812	2
12	4.9332792	1.893451911	8.7758E-05	1	1,054	1
13	5.3443858	1.955055641	2.89896E-05	0	1,340	0
14	5.7554924	2.013877247	9.25029E-06	0	1,674	0
Annual wind potential for $v > 3$ m/s (kWh/year · m²)						176
Average power density (W/m²) = E/8,760 h/year						20

$k = 1.4$. Figure 8.4 shows the resulting curves of probability distribution of the wind speed and the resulting energy density in the interval of wind speeds 4–14 m/s.

8.5 Variation of the Wind Speed with the Height above Ground

Air is a viscous fluid and as such, the speed of the wind is subject to a gradient between a point near the ground, where the roughness of the ground will slow down the flow to null speed, and a point high enough to let airflow undisturbed. As a rule, speed increases with height. There are several models to calculate the dependency between the "undisturbed" speed (for instance, at 50 or 100 m above ground) and the effective speed at our turbine's

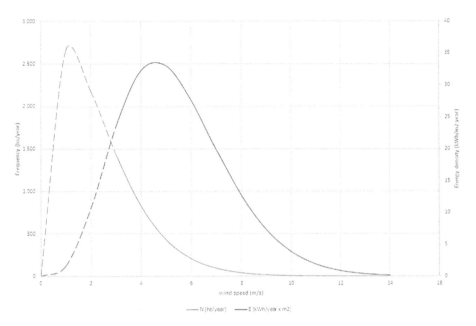

FIGURE 8.4
Weibull's probability distribution of the wind speed and of the energy productivity for the case shown in Table 8.3. The dotted portion of the curve represents the interval of wind speeds in which the turbine is not able to start. Observe that the maximum energy density of the site corresponds to $v \approx 4.5$ m/s, hence in this case the design of the turbine should be optimized for such speed.

hub, for instance, 15 or 20 m above ground. The classical formula that relates the wind speed and the quota above ground is called Hellmann's Law:

$$\frac{V_2}{V_1} = \left(\frac{H_2}{H_1}\right)^{\alpha}$$

Where:

   V_1 = wind speed at the highest quota under study (m/s)
   V_2 = wind speed at the lowest quota under study (m/s)
   H_1 = highest quota under study, usually 50, 100, or 200 m in the case of data from wind maps or from TMY provided by specialized companies
   H_2 = lowest quota under study, usually the foreseen installation height of the turbine's hub (m)
   α = land roughness coefficient, depends on the place, see Table 8.4.

Hellmann's Law is a model employed for simulations at high quotas and on large geographical scale, since it derives from theoretical considerations, assuming the Earth as a smooth sphere. It is acceptably accurate for interpolation, provided we know the speed at a high quota and can correctly estimate the exponent. On the contrary, if we know the speed from an anemometer

TABLE 8.4

Land Roughness Coefficient for Different Kinds of Geographical Areas

Ice or grass steppe	$\alpha = 0.08$–0.12
Flat land (sea, coastal area)	$\alpha = 0.14$
Prairie	$\alpha = 0.13$–0.16
Rural area	$\alpha = 0.2$
Steep land, forests	$\alpha = 0.2$–0.26
Very steep land, cities	$\alpha = 0.25/0.4$

placed at 10 m and need to extrapolate the speed at 50 m height, Hellmann's formula will lead to big extrapolation errors. The coefficients in Table 8.4 correspond to homogeneous "macroareas," i.e., in the order of kilometers around the point under study. When installing small wind turbines, we may face the situation in which the site is surrounded by small obstacles, lower than the turbine itself, but able to create some kind of interference anyway. The exponent in Hellmann's formula does not account for such local variations.

The most commonly employed height correction formula in the wind power industry is the logarithmic gradient formula, a mathematical model that works well for heights between 10 and 100 m.

$$\frac{V_1}{V_2} = \frac{\ln\left(h_1/h_0\right)}{\ln\left(h_2/h_0\right)}$$

Where:

V_1 = unknown wind speed at height h_1 (m/s)

V_2 = wind speed at height h_2, known from some data source (m/s)

h_1 = height at which we desire to calculate the wind speed (m)

h_2 = height at which the value V_2 has been measured or estimated (m)

h_0 = roughness length, defined as the height above ground at which the wind speed is null (m)

The value of h_0 can be easily estimated from Table 8.5, or when the "roughness class" of a site is known from wind maps developed on mesoscale models. The roughness class is a standard non-dimensional coefficient defined by the following formulas:

$$RC = 1.699823015 + \frac{\ln\ h_0}{\ln\ 150} \quad \text{for } h_0 \le 0.03$$

$$RC = 3.912489289 + \frac{\ln\ h_0}{\ln\ 3.3333} \quad \text{for } h_0 > 0.03$$

Roughness classes or roughness lengths are widely employed by the large wind turbine industry, and mapped in some publically accessible sites, like the **GWA**.

TABLE 8.5

Practical Values of h_0 and Their Correlation with the Roughness Class of a Site

Landscape Type	Roughness Class	Roughness Length h_0 (m)	Energy Index (%)
Water surface	0	0.0002	100
Completely open terrain with a smooth surface, e.g., concrete runways in airports, mowed grass, etc.	0.5	0.0024	73
Open agricultural area without fences and hedgerows and very scattered buildings. Only softly rounded hills	1	0.03	52
Agricultural land with some houses and 8 m tall sheltering hedgerows with a distance of approx. 1,250 m	1.5	0.055	45
Agricultural land with some houses and 8 m tall sheltering hedgerows with a distance of approx. 500 m	2	0.1	39
Agricultural land with many houses, shrubs and plants, or 8 m tall sheltering hedgerows with a distance of approx. 250 m	2.5	0.2	31
Villages, small towns, agricultural land with many or tall sheltering hedgerows, forests, and very rough and uneven terrain	3	0.4	24
Larger cities with tall buildings	3.5	0.8	18
Very large cities with tall buildings and skyscrapers	4	1.6	13

Source: Danish Wind Industry Association, http://xn--drmstrre-64ad.dk/wp-content/wind/miller/windpower%20web/en/stat/unitsw.htm#roughness, adapted by the Author.

The speed gradient can vary much with the *shape* of the obstacles. For instance, placing a wind turbine on top of a rounded hill will favorably increase the gradient, since the wind will be "concentrated" at the top. On the contrary, placing a wind turbine on top of a high building, near to the roof, may prove a disaster, because the turbulence induced by the edges of the building will result in a reduced average speed across the rotor, or even worse, negative speed gradients will induce noxious vibrations.

The speed gradient varies with the atmospheric stability too. The atmosphere is said to be unstable when vertical ascending currents are created by local heating phenomena, for instance, on large asphalted parking areas. When the atmosphere is unstable, the wind gradient diminishes, i.e., the wind speed is less dependent on the height. The atmosphere is said to be stable when descending thermal currents appear as a consequence of cooling of the site. In such condition the wind speed gradient increases much with the height. The atmosphere is neutral when there are no vertical air currents. Hellman's formula and the logarithmic gradient formula are valid only under neutral atmospheric conditions.

8.6 Practical Exercise N. 1

We desire to install a small horizontal axis wind turbine on the roof of a beach bar in Lignano, Italy (coordinates 45°39′37.5″N, 13°06′27.6″E determined with Google Maps). The foreseen 3-blade turbine has $D = 4\,m$, $C_{Pmax} = 0.39$ in the range 10–11 m/s, $V_{cut-in} = 3\,m/s$, $V_{cut-off} = 14\,m/s$, $P_{nominal} = 3\,kW$ at 10 m/s, $P_{max} = 4.2\,kW$ at 12 m/s. The blade's pitch is fixed, the speed is limited electronically.

We are going to check the annual electricity productivity with Weibull's and Rayleigh's functions and with data from the local weather station.

8.6.1 Estimating the Energy Productivity with the Help of a Wind Map and Weibull's Function

As a first step, visit the **GWA** http://globalwindatlas.com/map.html. Follow the tutorials to get familiar with its operation. Enter the coordinates of the place: Latitude = 45.660825 N, Longitude = 13.107848 E. Press **Go** and then **Mark**. A dot will indicate the position chosen. Now, in the left menu, click on **Direct 1,000 m Sampled Results**, and check **50 m** (this means the average wind speed at 50 m height). The result is shown in Figure 8.5, from where we read $V = 5\,m/s$.

Please note that the mesoscale model divides the territory in square grids, and the point under study falls within two areas with homogeneous mean speed, 4 m/s at the W and N of the point and 5 m/s at E and S. Hence, it is more accurate to assume 4.5 m/s as a more probable value for the point under study.

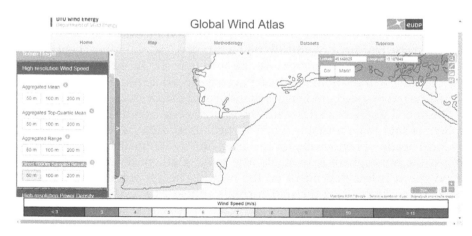

FIGURE 8.5
Average wind speed at 50 m height calculated on a 1 km mesh.

Now we must estimate the coefficients k and c. The GWA provides an information that helps to calculate both: the site's power density. Select **High resolution power density** in the left menu and then click on the button **50 m** under **Direct 1,000 m sampled results** (see Figure 8.6).

With the same criterion already explained, since the point is in a cell between two homogeneous areas of different power density, it is more reasonable to adopt a mean value, $162\,W/m^2$.

In the file *statistical-functions.xlsx* attached to this book, the sheet *estimation of k and c*, is self-commenting. The first step is to input the average wind speed V at 50 m height and the power density, respectively, $4.5\,m/s$ and $162\,W/m^2$, as estimated if the paragraph above. The process of finding the most accurate couple of values for k and c is iterative. At the third iteration, we obtain $k = 1.433$ and $c = 4.954$.

Please also note that in calculating the power density, we assumed the "standard" air density at $15°C$ and at sea level, $1.22\,kg/m^3$, because such is the standard adopted by the GWA and the wind power industry in general.

Suppose that the wind turbine's hub will be placed at 20 m above ground. The **GWA** can display the mapping of the local roughness height h_0. In the left menu **Land Use and Roughness**, choose the option **GWA Roughness**. With such information, it is possible to estimate the average wind speed at the desired height by means of the logarithmic law. Figure 8.7 shows a zoomed screenshot of the site.

Observe that the site is between two discrete areas with diametrically opposite roughness. Although formally the roughness height of the chosen coordinates corresponds to 0 because the selected site is on a shore, the latter is adjacent to urban buildings in a touristic city, which corresponds to roughness 1. The GWA samples on $1 \times 1\,km^2$ grids and interpolates to $250 \times 250\,m^2$ grids, so common sense says that in a case where the cells adjacent to the one of the chosen site have such diametrically opposite roughness

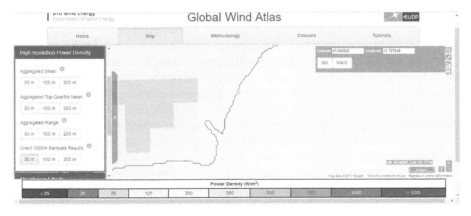

FIGURE 8.6
Geographical distribution of the power density of the site.

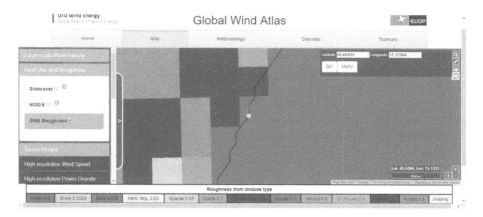

FIGURE 8.7
Map of the local roughness height, h_0.

factors, we should adopt a more conservative average value of h_0, hence in our calculations we will take $h_0 = 0.8\,\text{m}$. The reader can check with Google's satellite view that the site is effectively on a narrow shore, very close to a building and high trees, so the assumed value of the roughness length is a reasonable supposition, considering that the point is nearer to the buildings and trees than to the sea.

Now enter the value of k from the sheet *estimation of k*, and the values of V and h_0 estimated from the GWA, in the sheet *Weibull*.

Enter the same average speed in the sheet *Rayleigh*, in order to compare the accuracy of both formulas with the real data analysis presented in the next section.

8.6.2 Calculating the Energy Productivity with Anemometric Data Provided by the Local Meteorological Service

The local meteorological service provided the anemometric data of 2009 only. Such dataset cannot be considered a TMY, since a "true TMY" must be generated on the basis of data gathered along at least a decade. The year 2009 could have been an extraordinarily windy year, or the contrary, so we do not know if such data can be considered really representative of the local wind potential in our site. For sure, data of a complete year's measures from a near station is <u>more likely</u> to represent the local potential better than the simulations performed with wind maps and statistical functions, but it is subject to the measurement uncertainty estimated in Section 8.3.3.5. We highlighted "more likely" because the extrapolation of real data in a given site to a different site several kilometers away is not necessarily better than deriving a statistical distribution from mesoscale models. The ideal situation is to have both sets of information and combine data, a situation we will analyze in the next Section 8.6.2.

In this study case, the weather station is some kilometers away from the place where we envisaged installing the turbine, and at the time of performing the feasibility study (2010) the GWA was not functional. So we can assume that the land's roughness is comparable for both sites, an assumption that is fairly well confirmed by the satellite view. We will then work on the available data with the help of a standard spreadsheet. The file *lignano. xls* included with this book shows the procedure step by step. The first sheet, called *raw data*, shows the data "as provided" by the local meteorological station. The file's header contains the location of the sensor: latitude, longitude, and altitude. When the altitude is omitted, assume 10 m above ground, since this is the standard height of anemometric poles. In this particular case, the sensor is placed at 15 m height, as stated in the file's heading. Please note that in this particular case the anemometer logs vector averages every 30 min. We have two alternatives to handle this matter: assume that the result will be underestimated, or increase by 10% of all values, which will give a result closer to the scalar average, but with more uncertainty than that resulting from scalar data logs. In this example, we chose to increase the speeds by 10%. It is always good practice to make a copy and work on it, in order to preserve the original data. In the copy, called *find wrong data*, our first task is to search for the usual errors that occur when downloading and converting data from the weather station's software to commercial spreadsheets. Using the function "search" of your spreadsheet, look for "o" instead of "0", or ";" instead of ".", or blank cells, or special characters like "-" or "*". In this example, the original data provided by the local weather service contained the alphabetic strings "---" in cells C801, C9091, and C9395. You can either replace them with 0 or interpolate linearly between the adjacent cells, the goal is to have a set of purely numerical data otherwise the spreadsheet will give an error message in the next steps. In the next sheet, called *clean data*, we increased by 18% of the original vector average speeds: 10% in order to simulate scalar average wind speed from the available vector data and 8% because the data were sampled at 15 m height, but our turbine will be installed at 20 m height. The speed increment with height was obtained with the logarithmic law, assuming $h_0 = 0.80$ because of the considerations already explained in Section 8.6.1. The next step is to create a cumulative histogram. Please note that spreadsheets can create either Pareto histograms (ordinated) or cumulative histograms (a distribution by classes similar to Weibull's and Rayleigh's functions). We need to choose the second option, and the result will be the sheet named *Histogram* in our example file. Please refer to your spreadsheet's user manual on the necessary procedure, since this may vary from one software version to the other. Since the dataset contains the semi-hourly speed averages, we need to divide the frequency values in the sheet *Histogram* by two, in order to obtain a distribution that can be compared to those already calculated in Section 8.6.1 with Weibull's and Rayleigh's functions, whose measure unit is "hours/year." The result is shown in the foil *Energy analysis*.

8.6.3 When Both Mesoscale and Anemometric Data Are Available

Good practice in engineering cannot rely only on maps and online resources: visiting the site helps finding its peculiarities. The photo in Figure 8.8 shows the trees that grow immediately after the beach bar where the turbine was to be installed. Conifers are interesting as biological indicators, because they present the same exposed are to the wind throughout the year. In places with dominant winds, the trees suffer permanent deformation. Some researchers have found a direct correlation between such deformations and the average speed and duration of the winds in a site, though the results are not trustworthy and not applicable to all species of conifers. We can anyway extract an important information from the photo: the place has dominant winds blowing from the sea to the land, strong and frequent enough to deform the first rows of trees of the adjacent forest, but not enough to bend the successive rows.

On the other hand, the file provided by the local weather station contains wrong coordinates: LAT = 45.4102 N, LON = 13.0906 correspond to a point in the middle of the Adriatic. Asking the provider of the data, it was possible to find out that the actual coordinates are LAT = 45.69806 N, LON = 13.14361 E. Now observe Figure 8.9 that shows the place where the anemometer is installed: the local marina, which is behind residential buildings taller than the anemometer's pole. The marina is not adjacent to the sea but in a lagoon west of the town, i.e., on the opposite side of the town and sheltered from the dominant E-S.E. winds by the taller buildings.

The geographical differences between both sites provide us an important clue: the average wind speed and resulting power density measured at the

FIGURE 8.8
Trees adjacent to the place intended for the installation of the turbine. Observe how the first rows are bent toward land, while the next trees are straight.

FIGURE 8.9
The place where the anemometer is installed, seen from the direction of the dominant winds, behind the taller buildings and trees in first plane.

sheltered weather station are not directly applicable to the site on the exposed beach, where the average wind speed is more likely those resulting from the mesoscale model at 50 m. The data measured at the weather station cannot be employed to obtain empirical values of k and c to produce a Weibull distribution valid for the site under study, even if the distance between both sites is just a few kilometers. Please note that, according to the mesoscale model, the shape of the probability distribution would be the same for both sites, since at 50 and 100 m height—far above the average buildings and trees in the area—the average wind speed and energy density are the same for both the weather station and the beach sites. The difference between both sites at 20 m height is in their corresponding average speeds and power density, resulting from the different roughness and strongly influenced by the exposure to the dominant winds.

8.6.4 Conclusions

In Section 8.3.3.1, we determined that the measurement error—or uncertainty—of the power density measured with standard meteorological anemometers, and the assumption of constant air density, is in the order of 30%. Comparing the result from actual measures in the foil *Energy analysis*, 755 kWh/m²·year, and the result of the foil *Weibull* based on mesoscale data, 667 kWh/m²·year, the difference between them is 13%, i.e., well within the estimated uncertainty range. In principle, we can conclude that the quality of the results obtained with both methods is comparable, although the correlation between measured data and mesoscale data is not direct. In this particular study case, the data obtained with the mesoscale model are *more likely* applicable to the site than those measured directly in the weather station.

The result obtained with Rayleigh's probability distribution is too different from the other two. This is a practical confirmation that Rayleigh's function is not accurate when modeling sites with weak, unsteady winds. Table 8.6 and Figure 8.10 show the comparison between the results obtained with the procedures described in the former sections.

Since the energy density is proportional to the cube of wind speed, its probabilistic distribution resulting from the former date will be a curve whose peak will never coincide with the average speed of the place. Figure 8.11

TABLE 8.6

Comparison between the Results Obtained from Statistical Functions, and a Full Year Dataset Assumed as "TMY"

	Weibull (GWA)	Rayleigh (GWA)	Year 2009 (Weather Station Data)
Reference height for V (m)	50	25	15
V at 20 m (m/s)	3.5	3.5	2.9
E at 20 m (kWh/m²·y)	667	439	755
Productivity with a commercial turbine, 3 m/s $< v <$ 14 m/s (kWh/year)	2,663	2,067	3,490

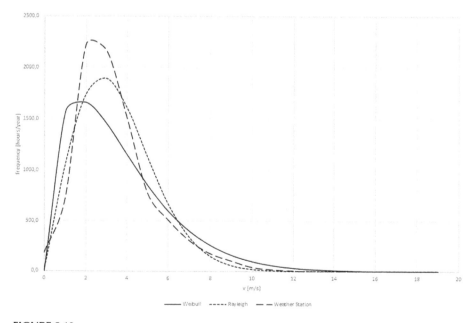

FIGURE 8.10
Comparison between the results from statistical functions and wind speed data measured in a location close to the site.

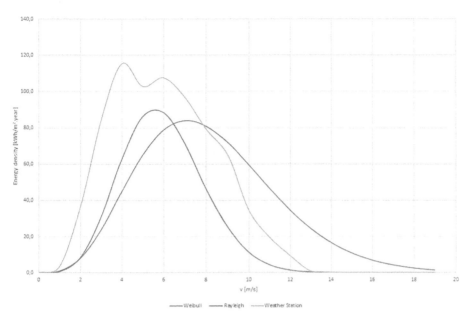

FIGURE 8.11
Probability distribution of the energy density calculated with statistical functions and with data from the local weather station.

shows the comparison between the energy density resulting from the Weibull and Rayleigh functions and weather station data.

From this example of a real study case, we can drive the following general conclusions, applicable to most cases:

a. Average wind speeds at high quota, when reduced to the height of the wind turbine's hub, are susceptible to overestimation. The logarithmic gradient function gives good approximation to the reality, on condition that the choice of the h_0 coefficient from the roughness map and the most probable V from the speed map is done correctly.

b. The turbine's rotor should be optimized for the wind speed that provides the maximum energy productivity, which will be higher than the average speed, because the energy is proportional to the wind speed's cube. Hence, it is always necessary to plot the curve of energy productivity as a function of the wind speed's probability, in order to find the peak of the energy density curve, and the corresponding wind speed that produces most of the energy during the year.

c. From a monetary point of view, the matching between curves is a second-order problem for the siting of small wind turbines. Compare, for example, the total net energy obtained with an actual turbine, when employing the Weibull distribution (2,663 kWh/year) and data from real measurements (3,490 kWh/year). The total difference between the productivity resulting from both models is 831 kWh/year. It may seem a lot, but if we consider the best feed-in tariff in Italy (in 2017) is 0.30 €, the financial uncertainty is just 249 €/year. Purchasing a set of 10 years validated data from the Italian meteorological authority (Military Aeronautics) would have costed nearly 700 €, and would not reduce the uncertainty. Buying mesoscale data with finer grid resolution from a specialized company would cost at least 2,000 €. For a small wind power project, as the one proposed in this example, wind data from free-access sources, and a bit of common sense in evaluating the roughness coefficient, are more than enough for obtaining acceptable results.

   d. We have demonstrated that the uncertainty in the determination of the energy density is ±30%. For practical purposes, when data from actual measures in a near location and from mesoscale models are available, we can split the uncertainty. In our example, the energy density from Weibull's formula +30% gives 3,462 kWh/year, and conversely, the energy density from the weather station's measures −30% gives 2,443 kWh/year. The intersection of both uncertainty ranges is a new range between 2,663 and 3,462 kWh/year. Assuming 3,062 kWh/year as design value of the installation will minimize the uncertainty of the energy production to a range of ±120 €/year.

8.7 Practical Exercise N. 2

Most manufacturers of small wind turbines propose models that reach their nominal power at 10 m/s wind speed. Why? Which consequences have such specification when installing the turbine in lower class sites?

8.7.1 Overrating: A Common Practice in the Wind Power Industry

The standard *IEC 61400-1 Design requirements* defines four "wind classes" and leaves a fifth class free for special designs, as summarized in Table 8.7.

The reference speed is defined as "the maximum average wind speed at the hub's height within a 10 min lapse that can be encountered in the site in a return period of 50 years." It is the value employed for the structural calculations.

The density of air according to the said standard is 1.225 kg/m³.

The Rayleigh probability distribution function is the standard adopted for comparisons, not only because it describes well windy places like Class

TABLE 8.7

Summary of the IEC 61400-1 Wind Classes

	Reference Speed (m/s)	Average Annual Speed (m/s)
Class I	50	10
Class II	42.5	8.5
Class III	37.5	7.5
Class IV	30	6
Class S	Values defined by the manufacturer	

I– Class IV, but also because its parameter is univocally related to the annual average speed *V*.

The said standard has its roots in the industry of large wind turbines, whose scope is energy production on massive scale, based on multi-million investments. It is clear that on such investment levels, sites rated lower than Class IV would offer uninteresting return of investment (ROI). Such vision has introduced a strong bias in the market of small wind turbines, although the logic of a potential small-scale user is not necessarily related to the ROI, but to non-economical parameters—e.g., self-sufficiency in remote areas, easiness of maintenance with local resources, turbine required for pumping water, or in the worst case, having no other viable alternative. This means that a user may need to install a small wind turbine even in a place with average wind speed by far lower than Class IV, and the ROI may still be acceptable, or not the decision factor.

Small wind turbine manufacturers tend to design their products for Class I sites for marketing reasons: using measures of power in describing wind turbines turns in the benefit of the manufacturer because it is easy to abuse. Technically, wind turbines can be designed with high power ratings relative to their swept area for very windy sites. Nevertheless, some manufacturers have played on this and marketed wind turbines with very large generator ratings relative to the turbine's swept area. Why? Because unsophisticated buyers often compare wind turbines on their installed cost relative to their installed capacity and not on their cost relative to their effective productivity. By inflating the wind turbine's rating, the manufacturer can charge more money than a competitor and still look cheaper in $/kW of installed capacity, just because the rotor will be smaller, the pole will be consequently shorter, and the foundations proportionally smaller. Modern mass production in China of niobium supermagnets and high efficient inverters has also made alternators and ancillary equipment cheaper, so the current production cost is mainly related to the diameter of the rotor, not to the rated power. According to the British association Green Spec, hyped claims of power output by small wind turbine manufacturers is probably the cause of most of the unhappy experiences of household wind power generation in the UK, http://www.greenspec.co.uk/building-design/small-wind-turbines/?$.

As a rule for the correct design of a small wind power system, the wind speed corresponding to the rated power **should not** coincide with the average wind speed of the site, nor with the highest IEC Class. An optimum turbine must provide nominal power at, or near to, the wind speed corresponding to the peak of the local energy density distribution, as resulting from Rayleigh's or Weibull's formula or from local wind speed measurements.

8.7.2 Practical Example

Suppose we need to supply 3,000 kWh/year to a home in a remote rural area, where the average wind speed is 3 m/s. For market availability reasons the owner desires to install a wind turbine of a given brand, whose catalog and declared output power of the different models can be summarized in the following Table 8.8. Are any of these turbines suitable for the user's scope?

The sheet *Class I* in the file *statistical-functions.xlsx* allows to check which model is more suitable for the real case proposed, and how a generic turbine designed on purpose for the place compares to them.

8.7.3 Conclusions from the Example

The comparison shows that the only commercial model able to provide more than 3,000 kWh/year is the one rated 10 kW. Such model is three times bigger than necessary and more than five times overrated, since the average power it would produce in the installation site is just 1.7 kW. Please note that generators and inverters are efficient only when they are loaded at 80% or more their nominal capacity. Therefore, the standard 10 kW turbine may cost less than others when compared with the cost/power ratio, but it is absolutely uneconomical for the required mission. On the other hand, a generic wind

TABLE 8.8

Summary of the Available Models of the Brand Desired by the Customer

Model	3 kW	5 kW	10 kW
Area	11.34 m²	14.5 m²	75.4 m²
Wind Speed	Net Power Output		
2	0	0	0
5	0.3 kW	0.45 kW	2 kW
7.5	1.2 kW	1.5 kW	9 kW
10	2 kW	3.4 kW	10.1 kW
12.5	2.3 kW	4.3 kW	10.1 kW
15	2.5 kW	4.45 kW	10.1 kW
17.5	2.5 kW	4.5 kW	10.1 kW
20	2.5 kW	4.4 kW	10.1 kW
Rated wind class	I	I	I

turbine designed on purpose for the place would require a bigger rotor than the Class I, 5 kW model, with half of its rated power. It may cost more when compared with cost/power ratio, but its absolute value will be nearly half that of the 10 kW, Class I model, because the generator is smaller, the mast and foundations can be sized for a lower reference speed, (requires some specific survey, most probably in the range 15–20 m/s), the inverter and battery bank will work most of the time at higher efficiency.

Bibliography

Arrojo C.D., Díaz J.G., Dampé J.C. et al., *Medidas eléctricas*, notes on the course on Electrical Measures, National University of La Plata, Argentina, 2013.

BIPM (Bureau International des Poids et Mesures), *Evaluation of Measurement Data. Guide to the Expression of Uncertainty in Measurement*, JCGM 100, 1st edition, BIPM, Sèvres, France, 2008.

Carrillo C., Cidrás J., Díaz-Dorado E., and Obando-Montaño A.F., An approach to determine the Weibull parameters for wind energy analysis: the case of galicia (Spain), Energies 7, 2676–2700, 2014. doi:10.3390/en7042676, Free to download from www.mdpi.com/ 1996-1073/7/4/2676/pdf.

ESMAP The World Bank's Energy Sector Management Assistance Program, *Best Practice Guidelines for Mesoscale Wind Mapping Projects for the World Bank*, October 2010. Available for free download http://www.esmap.org/sites/esmap.org/files/MesodocwithWBlogo.pdf.

Girma Dejene Nage, Analysis of wind speed distribution: comparative study of Weibull to Rayleigh probability density function; a case of two sites in Ethiopia, *American Journal of Modern Energy* 2 (3), 10–16, 2016. doi: 10.11648/j.ajme.20160203.11, Free to download from http://article.sciencepublishinggroup.com/html/10.11648.j.ajme.20160203.11.html.

IEC Standards, *IEC 61400-12-1:2017 Wind energy generation systems Part 1: Design requirements. Part 12-1: Power performance measurements of electricity producing wind turbines*.

Rosato M., *Diseño de máquinas eólicas de pequeña potencia*, Editorial Progensa, Sevilla, Spain, 1992.

Rosato M., *Progettazione di impianti minieolici*, multimedia course (in Italian), Acca Software, Montella (AV), 2010.

9

Sizing Energy Storage Systems

9.1 Stand-alone Wind Power Generators

The most recent regulations in many industrialized countries allow the connection of small wind turbines—or other renewable energy systems—in parallel with the grid. The operational advantage of the grid interconnection is the absolute flexibility for the user: when the wind produces more energy than needed, the excess is conveyed elsewhere by the grid, when there is no wind at all, the user gets the necessary energy from the grid. In such cases, there is no need for installing an energy storage system, and the cost of the installation is kept to the minimum. There are circumstances when it is necessary to store energy—not only electric energy, although this is the usual case—in order to adapt the wind availability to the user's demand. The most frequent case is that of rural users in places too far away from the public grid, where the cost of building a transmission line for serving a very small load would be prohibitive. There is a growing interest among certain citizen groups toward the energy independence, often because of "ideological" rather than economical reasons: "anti-system attitude," "preparing for the next global crisis," or just "being green." There are several energy storage techniques, for instance, as compressed air, pumping water to a higher tank or basin, heating water in an insulated tank, as electric or magnetic field, etc. Nevertheless, the most efficient way to accumulate energy currently in use is employing a battery bank and an inverter that converts the direct current (DC) into the more versatile alternating current (AC).

9.2 Stationary Batteries for Electrical Energy Storage

Batteries currently in commerce can be of three different kinds, depending on their scope: automobile starters, traction batteries, and stationary batteries. The first ones feature a high capacity to provide the big peak currents absorbed by automotive start motors, but in general are not able to maintain the current constant with time, and will become irreversibly damaged if completely discharged. Traction batteries were traditionally employed in forklifts and other industrial vehicles, as backup lighting supply in train coaches, airplanes, and ships, and more recently also as power supply in electric cars. They are able to provide strong currents during short-time intervals, tolerate cycles of complete charge and discharge, and have the best capacity/weight ratio, an important aspect related to the performance of electric vehicles, but irrelevant for the designer of static storage systems. Stationary batteries are mostly employed as backup energy supply in case of blackout for large buildings or critical applications (hospitals, communication systems, banks and other data processing facilities, coastal beacons, airports, etc.), or as energy storage in places not served by an electric grid. They feature the capacity to tolerate charge current even when they are already charged and are able to yield all the stored energy without damage. On the other hand, stationary batteries are not able to provide large surge of currents, and their capacity/weight ratio is the lowest in the market.

Many battery technologies exist, but at the current date (2017) the most common batteries are:

- The lead-acid type—mostly employed as car starters and stationary banks;
- The alkaline Ni–Cd, Fe–Ni, and their variants;
- The Li or Li-ion technology—mostly diffused in cell phones, portable PCs, and advanced electric vehicles and, finally;
- The Li-Phosphate battery, employed in Germany as residential stationary storage because of its highest safety features against overcharging and overloading.

In this chapter, we are going to show which parameters to consider when designing an electric storage system for a wind turbine.

9.2.1 The Charge–Discharge Capacity

Typically, the charge and discharge capacities are expressed in Ampère per hour ($A \cdot h$), because it is conventionally assumed that the voltage remains constant. Remember that in DC systems, the electrical power (W) is the product of the voltage by the current ($V \cdot A$); hence, the energy stored or released is the product of the power by time ($W \cdot s$ or $W \cdot h$). The capacity of charge or discharge is then measured in ($A \cdot h$), because the voltage will depend on how many batteries the system has, and how they are connected (series or parallel). In general, batteries subject to long cycles of charge and discharge will have a longer life and a higher storage capacity than the nominal. Consequently, battery manufacturers specify the nominal capacity of the battery with the letter C, followed by a suffix that represents the test's duration in hours. For instance, a battery labelled 300 A·h C_{100} is able to supply 300 A·h on condition that the duration of the discharge is 100 h; hence, during the said time the current must be 3 A or less. Discharging the battery at a higher rate, for instance 10 A, will result in a discharge duration shorter than the theoretical 30 h, hence, a smaller effective capacity.

9.2.2 The Discharge Depth

Each kind of electric battery admits a given discharge depth, which in turn is related to the useful life, which is measured in cycles of charge–discharge. The useful life of a lead-acid battery follows an exponential law of the type:

$$U.L. = A \cdot e^{(-s/k)}$$

where:

U.L. = useful life (total number of charge/discharge cycles that the battery can withstand)

A = maximum number of cycles that the battery can withstand under negligible discharge depth (8,000 for lead-gel stationary batteries, 10,000 for lead-acid stationary batteries; 4,000 for regular lead-gel automotive batteries, 5,500 for high capacity automotive lead-gel batteries)

s = discharge depth (% of the battery's nominal capacity)

k = factor that depends on the type of battery (0.3–0.35 for stationary lead batteries; 0.24–0.28 for high capacity automotive lead batteries)

The life of alkaline stationary batteries, on the contrary, is nearly independent of the discharge depth.

The standard to determine the maximum number of cycles that the battery can tolerate during its useful life is the IEC-896-2, which states that the cycle must last 3 h at the C_{10} discharge current, until the battery reaches 60% of its capacity. For instance, a battery labeled IEC-896-2 "600 cycles" will last more than 600 cycles if the depth of discharge does never exceed 60% of its nominal capacity. Most manufacturers provide additional information to the IEC-896-2 number of cycles, as for instance curves similar to the one shown in Figure 9.1.

As a rule, the shallower the discharge, the longer will be the useful life of the battery. On the other hand, the size and cost of the energy storage system will be bigger.

9.2.3 Self-Discharge Percentage

Any battery left open circuit instead of under polarization charge, will spontaneously discharge. The self-discharge percentage is the amount of charge that the battery will lose in a reference time lapse, usually 30 days.

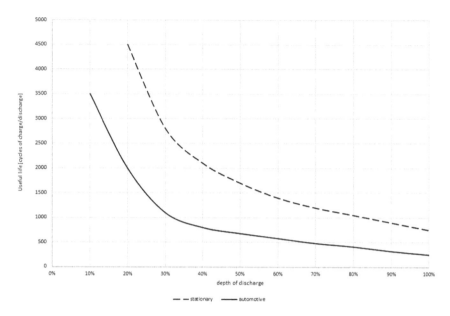

FIGURE 9.1
Comparison of the useful lives of two "maintenance free" batteries, as a function of their discharge depth. Data from the catalogs of two different manufacturers, graphics by the Author. *Legend*: Stationary: a lead-gel stationary model for photovoltaic applications. Automotive: a lead-acid battery with AGM technology for car motor starting.

9.2.4 Choosing the Most Suitable Battery for a Given Scope

We will not analyze the peculiarities of lithium batteries in this book, because at the present date (2017) their cost is too high for applications requiring big storage size. The price of lithium batteries may fall in the future because of scale economies induced by the car manufacturing industry, but for small wind power systems the only current alternatives for designing an off-grid energy storage are two: stationary lead-acid batteries and alkaline Ni–Fe and Ni–Cd batteries. Lead-acid batteries are in general cheaper and widely available, but their useful life depends strongly on the discharge depth. Hence, they need to be replaced periodically, every four to five years for daily cycles shallower than 60% of the rated capacity. Alkaline batteries are more expensive, but their useful life is almost independent of the discharge depth, in general more than 10,000 cycles, i.e., 27 years for daily cycles.

Table 9.1 shows the differences between both typologies.

Choosing one or the other type depends both on economical and logistic factors in each place. First, it is necessary to check the manual of the inverter envisaged for the installation, because some models have built-in charge regulators suitable only for lead-acid batteries.

TABLE 9.1

Comparison between Stationary Lead-Acid and Alkaline Batteries

Factor to be Considered	Alkaline Batteries	Lead-Acid Batteries
Useful life	>30 years	<10 years
Can remain discharged for long time	Yes	No
"Memory effect" (progressively losing the initial capacity because of charge/discharge cycles)	Yes in some old models of the Ni–Cd type; no in modern models	No
Needs voltage regulator	No	Yes
Voltage drop with strong current	No	Yes
Exhalation of vapors (wet batteries)	Emits H_2 and O_2 when charged over 80%	Corrosive vapors
Equalization charge (necessary to eliminate salt deposits on the electrodes, that reduce the capacity of the battery)	At least twice a year, charge at 110% of nominal V	Each month, charge at 110% of nominal V
Quick charge	6–7 h without problems	8–16 h
Disposal at end of life and toxicity	Environmental risk! Bring back to specialized recycling facility or to provider	Environmental risk! Bring back to specialized recycling facility or to provider
Energy density	30–50 W/kg	45–80 W/kg
Damage in case of overload or overcharge	No	Yes
Self-discharge (%)	10% the first 24 h, then 20% monthly	3%–5% monthly
Real capacity	Up to 100% of nominal capacity	Advisable to discharge less than 60% of the nominal capacity
Operating temperature	0°C to 45°C	–20°C to 50°C
Cell voltage (fully charged)	1.2 V	2 V
Minimum cell voltage (discharged)	1 V	1.75 V

Batteries belong to two categories: "wet" (a.k.a. "flooded") and "dry," the latter usually called "maintenance-free" or "sealed," regardless of them being alkaline or lead-acid.

Wet batteries must be installed in a well-ventilated room, in vertical position to avoid corrosive electrolyte spills, and require the periodical control of the level of electrolyte, eventually adding distilled water. Dry batteries usually contain the electrolyte in the form of gel, are sealed and do not require maintenance operations, nor do they exhale corrosive or explosive vapors. Their drawback is that they can explode if overloaded or overcharged, and it is impossible to prolong their useful life by extraordinary maintenance operations. Lead-gel batteries usually have lower performance than lead-acid models, because the gel increases the internal resistance, lowering the maximum current and efficiency. An intermediate technology is the valve sealed wet battery with absorbed glass mat (AGM). Such batteries have special valves that allow vapors to condense and precipitate back into the electrolyte container. Formally, they are considered sealed batteries because vapors cannot escape, but the valves can be removed for eventual extraordinary maintenance, which allows extending their useful life. Fiberglass mats placed between the plates allow the free circulation of the electrolyte but prevent short-circuiting.

9.2.5 Influence of Temperature and Discharge Rate

Lead-acid and lead-gel batteries can operate in the range −40°C to +50°C. Their capacity varies with the temperature as well as with the discharge rate, in the way shown in Figure 9.2.

9.3 Examples of Stand-alone Wind Power System Design

9.3.1 Feasibility of Using of Standard Automotive Batteries for Stationary Applications

Consider the case of a leisure home on a mountain, settled only 48 h during the weekends. The total energy demand when occupied is estimated in

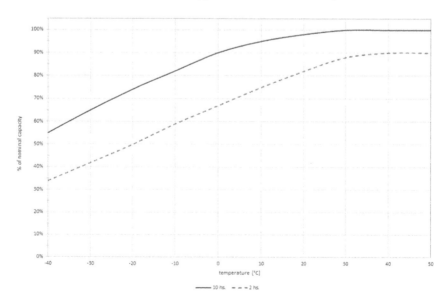

FIGURE 9.2
Variation of the capacity of a lead-acid battery with the temperature, for 10 and 2 h discharge rates.

8.5 kWh/week. The average temperature is 5°C. The maximum load (sum of the appliances switched-on simultaneously) is 850 W. The predefined inverter requires 48 V. The battery bank must be able to provide the following nominal and peak output currents:

$$I_{nom} = \frac{8500\,(Wh)}{48\,(h) \cdot 48\,(V)} = 3.7\,(A)$$

$$I_{peak} = \frac{850\,(W)}{48\,(V)} = 17.7\,(A)$$

Suppose we decide that 10 years is an acceptable useful life for our battery bank. Since there are 52 weekends in one year, the optimum useful life of the battery will be 520 cycles. According to the curves in Figure 9.1, we could discharge up to 70% of a bank composed of standard car batteries, or discharge 100% of the bank of stationary batteries but in the latter case, the batteries would provide more than 1,000 cycles of useful life, i.e., nearly 20 years lifespan. The decision is then purely economic and/or logistic: Is it better to install a smaller bank of expensive batteries that will last 20 years, or a bigger bank of cheap batteries that will last 10 years?

The only way to answer rationally such question is to check both cases.

9.3.1.1 Automotive Batteries

Assuming 70% discharge depth, the required effective capacity is:

$$C_{eff} = \frac{3.7\,(A) \cdot 48\,(h)}{0.7} = 254\,(A \cdot h)$$

The resulting C_{10} current that such battery can provide will be 25.4 A, more than the expected peak current, so we will not have any capacity reduction because of peak currents. On the other hand, the place's average temperature is lower than 20°C so, according to Figure 9.2, the effective capacity will be reduced to 95% of the nominal rating. Hence:

$$C_{nom} = \frac{C_{eff}}{0.95} = 267\,(A \cdot h)$$

We can now search for the closest model in the catalog of the manufacturer. In this case, the biggest standard model has 140 A·h, so we can slightly exceed the desired design capacity by connecting such batteries in parallel, obtaining then 280 A·h. The bank will be composed of two series of four batteries connected in parallel, as shown in Figure 9.3. The voltage of such array is $4 \times 12\,V = 48\,V$ and its capacity is $2 \times 140\,A \cdot h = 280\,A \cdot h$.

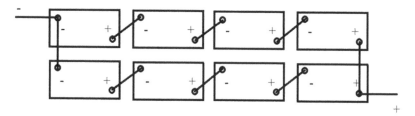

FIGURE 9.3
Parallel connection of two groups of four batteries connected in series.

9.3.1.2 Stationary Batteries

Assuming 100% discharge depth, the required effective capacity is:

$$C_{eff} = \frac{3.7(A) \cdot 48(h)}{1} = 178(A \cdot h)$$

The resulting C_{10} current that such battery can provide will be 17.8 A, nearly the expected peak current, so we will not have any capacity reduction because of peak currents. On the other hand, the place's average temperature is lower than 20°C so, according to Figure 9.2, the effective capacity will be reduced to 95% of the nominal rating. Hence:

$$C_{nom} = \frac{C_{eff}}{0.95} = 187(A \cdot h)$$

We can now search for the closest model in the catalog of the manufacturer. In this case, the closest standard model has 200 A·h, so we can slightly exceed the desired design capacity. The bank will be composed of just four batteries connected in series.

9.3.1.3 Selection Factors

At this point, it is possible to compare the purchase cost of both solutions and their duration. Since the duration of the stationary bank will be 20 years, a correct comparison model should consider the said lifespan, and purchasing a new car battery bank after 10 years. Such analysis is easy to carry out with a spreadsheet, using the built in function Net Present Value (NPV) and defining the same interest rate for both cases. The file *battery-economy.xlsx* included with this book shows an example. The prices of the batteries are not real, just for didactic purposes. Table 9.2 shows the result. Observe that, in order to be economically convenient, the single car battery should cost less than 27% of the single stationary battery.

Another disadvantage of series/parallel arrays is that the number of connections is double; hence, the resistive losses caused by contact resistance are bigger.

TABLE 9.2

Economical Evaluation of Both Alternatives

	Car Battery	Stationary Battery
Cost of the Single Battery	USD 27	USD 100
Number of Batteries	8	4
Interest Rate	2%	
Year	Total Cost	Total Cost
0	USD 216	USD 400
1	0	0
2	0	0
3	0	0
4	0	0
5	0	0
6	0	0
7	0	0
8	0	0
9	0	0
10	USD 216	0
11	0	0
12	0	0
13	0	0
14	0	0
15	0	0
16	0	0
17	0	0
18	0	0
19	0	0
20	0	0
Net present value	USD 385	USD 392

9.3.1.4 Size of the Wind Turbine

Having defined the necessary weekly energy and the size and kind of the battery bank, we can now calculate the size of the turbine. We must either know the local average wind speed and apply one of the probability distribution functions explained in Chapter 8, or have enough data measured on site with an anemometer, and find out the month with the weakest winds. Suppose that the lowest local average wind speed is 3.5 m/s, and applying the Weibull distribution formula to the operating speed interval of the turbine, 4–14 m/s, the resulting density of energy is $42.5 \text{ kWh/m}^2 \cdot \text{month}$ in the worst month. Suppose that the turbine will be of the horizontal axis type, having an estimated C_p equal to 0.38 (in this case it is intended as global C_P, i.e., the overall conversion factor including the efficiency of the alternator, rectifier, and voltage regulator). The energy available weekly at the terminals of the battery bank will be:

$$E = 8.5(\text{kWh/week}) = 0.38 \cdot 42.5(\text{kWh/m}^2 \cdot \text{month}) \cdot \frac{7(\text{d})}{30(\text{d})} \cdot S$$

$$= 3.8(\text{kWh/m}^2 \cdot \text{week}) \cdot S$$

In order to store in one week the energy required during the weekends, the minimum area of the turbine must be:

$$S = \frac{8.5(\text{kWh/week})}{3.8(\text{kWh/m}^2 \cdot \text{week})} = 2.24(\text{m}^2) \rightarrow D = 1.7(\text{m})$$

9.3.2 Example of Stand-alone Wind Power System Design with Stationary Batteries

Consider the case of a rural family in a remote area. Their total energy demand is estimated in 450 kWh/month, i.e. an average of 15 kWh/day. The maximum acceptable load (sum of the appliances switched-on simultaneously) is 3 kW. The predefined inverter requires 120 V DC. The battery bank must then be able to provide the following average and peak output currents:

$$I_{\text{nom}} = \frac{15,000(\text{Wh})}{24(\text{h}) \cdot 120(\text{V})} = 5.2(\text{A})$$

$$I_{\text{peak}} = \frac{3,000(\text{W})}{110(\text{V})} = 27.3(\text{A})$$

From the analysis of wind data, the lowest local average wind speed is 3.5 m/s, and applying the Weibull distribution formula to the operating speed interval of the turbine, 4–14 m/s, the resulting density of energy is 42.5 kWh/m^2·month in the worst month. The rest of the year, the average density of energy is 50 kWh/m^2. Historical series show that the wind speed will be lower than the cut-in speed for 24 consecutive hours only once a month, but in the worst month, historical series show that it is possible that such condition can last 48 h, once in the month. This means that the system must have the capability of supplying up to 30 kWh without recharge, in order to ensure continuity in the worst two days of the year. The depth of discharge will be 100% once in a year and 50% 11 times in a year. The rest of the time, it will be always smaller than 50%. In this case, there is no doubt that only stationary batteries can do the job because the installed capacity will be large and at least twice a year the system will be required to reach 100% depth of discharge. Sizing the battery bank can result from two different criteria: maximum durability or minimum size of the bank.

9.3.2.1 Maximum Durability Criterion

If we define arbitrarily that the acceptable lifespan of the battery bank is 10 years, the total number of cycles will be $365 \times 10 = 3{,}650$. According to Figure 9.1, the maximum advisable discharge depth to ensure such number of cycles is then 25%; hence, the nominal capacity of the whole battery bank is simply:

$$C_{nom} = \frac{5.2(A) \cdot 24(h)}{0.25} = 500(A \cdot h)$$

9.3.2.2 Minimum Size Criterion

From the considerations on the wind distribution of the site, the average depth of discharge in one year will be less than 50%. Assume it is 40%, hence

$$C_{nom} = \frac{5.2(A) \cdot 24(h)}{0.4} = 312(A \cdot h)$$

From Figure 9.1, the duration of the battery will be then 2,100 cycles, which corresponds to $2{,}100/365 = 5.75$ years. For one criterion or the other, once again, is monetary. The reader can check by himself/herself using the same file *battery-economy.xlsx* provided. Assuming the same cost per ampere hour for both alternatives, the criterion of maximum durability proves to be the most economic.

9.3.2.3 Worst Month

Having calculated the necessary capacity for maximum durability of the battery bank, we will now check if the same is enough for the worst condition of the year, two consecutive days without charge:

$$DoD = \frac{5.2(A) \cdot 48(h)}{500\,A \cdot h} = 50\%$$

The criterion of maximum durability not only fulfills the required autonomy in the worst days of the year, but also ensures a maximum extraordinary autonomy of four days without recharge.

9.3.2.4 Size of the Turbine

Now that the necessary weekly energy and the size of the battery bank have been defined, we can calculate the size of the turbine. Since the density of energy is $42.5 \text{ kWh/m}^2 \cdot$ month in the worst month, and the wind will blow with useful intensity for only 28 days then the available average energy in such days will be:

$$E_{net} = \frac{42.5(\text{kWh/m}^2 \cdot \text{month})}{28(\text{days/month})} = 1.52(\text{kWh/m}^2 \cdot \text{day})$$

Suppose that the turbine will be of the horizontal axis type, having an estimated C_p equal to 0.38 (in this case, it is intended as global C_p, i.e., the overall conversion factor including the efficiency of the alternator, rectifier, and voltage regulator). The generator must be able to produce 30 kWh in one day in order to accumulate enough energy for the eventual two days with no wind. Hence, the size of the turbine results from the following equation:

$$E = 30 \text{ kWh} = 0.38 \cdot 1.52(\text{kWh/m}^2 \cdot \text{day}) \cdot 1 \text{ day} \cdot S$$

$$S = 52 \text{ m}^2 \rightarrow D = 8 \text{ m}$$

Bibliography

Exide Technologies, *Handbook for Gel-VRLA-Batteries*, Technical Report Rev. 5, Büdingen, 2003.

North Star, *Pure Lead AGM, Automotive and Marine Application Manual*, North Star, Springfield, IL, 2016.

Rosato M., *Diseño de máquinas eólicas de pequeña potencia*, Editorial Progensa, Seville, 1992.

Rosato M., *Progettazione di impianti minieolici*, Multimedia course, Acca Software, Montella, 2010.

10

Design of Wind Pumping Systems

10.1 Water and Energy

Pumping water with wind power is an old and well-studied technique. Nowadays, the most diffused wind turbines for water pumping are the American windmill, self-made *Savonius* rotors, and sail mills (a.k.a. Cretan or Mediterranean windmill). Slow wind turbines are ideal to drive positive displacement pumps, i.e., piston, diaphragm, screw, peristaltic, or lobe pumps, because of their high starting torque. Positive displacement pumps have many advantages for wind pumping systems: robustness, low operational speed and hence low wear and maintenance cost, low capital cost, easy reparation with simple tools, high efficiency—especially for piston and diaphragm pumps—and finally, the capacity to pump from deep wells.

Pumping water with high speed turbines (vertical or horizontal) coupled to centrifugal or screw pumps is technically possible too. The overall efficiency of such systems is good, but their cost is higher. The market trend shows an increasing offer of preassembled systems, composed of a fast wind turbine with electric generator, battery bank, inverter, and conventional centrifugal pump—usually a submerged well pump—controlled by the same inverter. Such configuration allows a more flexible use of energy both for pumping and other scopes, but its pumping efficiency may be lower than direct-driven wind pumps. Choosing the most adequate technology for each case is not an easy task, as we will demonstrate in Section 10.4.

It turns spontaneous to ask oneself if the current market trend has a technical reason, or just follows the convenience of small turbine manufacturers. In short: Is it more convenient to store electricity in a battery bank, and then pump water when needed or just to store water in a tank during the windy season? Water storage in a tank or reservoir above ground is a way of storing energy; in many cases, cheaper than a battery bank, or at least easier to build in remote rural areas. Water storage could have some advantages in industrialized countries too. For instance, installing a wind turbine for electricity generation in Italy is subject to a series of bureaucratic authorizations and furthermore the generator and all electrical components must carry national and European labeling, which adds extra costs. On the contrary, a windmill for water pumping and its storage tank are not subject to the said requirements, and the procedures for the necessary installation permits are simpler.

When seasonal variations are such that the period of maximum water demand corresponds to that of minimum average wind speed, pumping water during the windy season and storing it in plastic membrane containers—a.k.a. *flexybags* or *flexytanks*, or *water bladders*—becomes economically interesting. Such solution requires neither civil works, nor construction permits. Another aspect to consider is the quality and reliability of the water supply. Global climate change as well as growing population is increasing the pressure on aquifers while desertification is advancing on fertile land. In some regions, the level of the water table sinks—or its salinity may rise—during the dry season, making the water unpotable for people and animals and useless for agriculture. In the worst case, the well may eventually dry out precisely in the period when water is more necessary. The dry season usually corresponds to the lowest wind speed season. Hence, pumping water during the windy season and storing it in adequate closed containers like *flexytanks*, to avoid evaporation losses, not only will require less energy because of the higher level of the water table, but it will provide better water quality and higher supply reliability too.

Water pumping for agricultural use can belong to one of two general categories:

a. Water for animal and human consumption.

   In this case, clean water is pumped from deep or very deep boreholes; the average flow is relatively modest and more or less constant during the year.

b. Water for crop irrigation.

   In this case, water is usually pumped from shallow wells, canals, rivers, or from one rice paddy to another. Such water is often muddy or contains suspended solids; the required daily flows are big during a limited number of weeks of the year, while the pumping head is low.

Such different requirements reflect in the choice of the pumps and windmills employed for one scope or the other. The first case is the most common situation in countries like the United States, where the classical water mill was invented around 1860, then spread to Argentina and Australia, and practically has not changed until now. In Argentina, the common solution employed in cattle farms is the "American windmill" pumping water into an "Australian tank." Examples of the second case are traditional pumps employed in Northern Africa, Middle, and Far East, where rope pumps (a.k.a. chain and washer pumps), waterwheels, Archimedes screws, and similar low-head pumps are driven either by animals or by simple sail or vane mills.

10.2 Water Pumps and Wind Turbines

10.2.1 Generalities

Water pumps are of two different kinds: centrifugal or positive displacement. The first ones are divided into slow (big diameter, low rotational speed) and fast (small diameter, high rotational speed) centrifugal pumps. The second ones are classified in reciprocating and rotating positive displacement pumps. Each kind of pump has advantages and disadvantages, as well as different energy efficiencies. In a very broad sense:

a. Slow centrifugal pumps are potentially suitable for low head, high flow rate applications, combined with vertical axis wind turbines, but their efficiency is good only in a very narrow range of rotational speed.

b. Fast centrifugal pumps require high rotational speeds that wind turbines cannot attain, so they are usually driven by electric motors, and their natural integration is in electric wind turbine systems.

c. Positive displacement pumps, either rotating or reciprocating, are suitable for self-construction, and their efficiency in general is more constant throughout a wider range of operational speeds, so they result potentially more suitable for the operation in wind-driven systems.

In this chapter, we will focus mainly on case c) because positive displacement pumps can potentially provide the most efficient solutions in different situations, either in developing or in industrialized countries. Centrifugal pumps will be treated superficially just for completeness.

10.2.2 Centrifugal Pumps

The flow of centrifugal pumps is proportional to the square of the rotation speed, and hence their absorbed power follows the same law, because the power demand of a pump is the product of its flow rate by its water head. Single-stage centrifugal pumps generate a limited pressure, so they are more suitable for low heads and high flow rates. When bigger heads are required, their design and cost increase because of the higher complexity of multistage rotors. The suction head of centrifugal pumps is practically null, so they need to be manually primed (suction heads <6 m), or their rotor must be installed below the water level (submerged pump or sump pump), or a Venturi ejector must be placed close to the water level. Submerged pumps are energetically more efficient than those with Venturi ejector are, but their cost increases with the depth of the borehole because a bigger number of rotors in series will be necessary to overcome the water head, and a big diameter borehole is required to accommodate both rotor and shaft. There are two possibilities to power centrifugal pumps with a wind turbine:

a. Using a standard electrical submerged pump. In such case, it is necessary to install an electrical wind turbine, a small battery bank to stabilize the voltage and an inverter to provide the necessary constant speed of rotation;

b. Using a standard submerged centrifugal pump with direct shaft coupling. Such solution is theoretically interesting when low heads and high flow rates are required, or when such kind of pump is already in operation, and the scope of the wind turbine is to replace an existing petrol motor. Figure 10.1 shows the conceptual adaptation of a vertical axis wind turbine to an existing shaft-driven submerged borehole pump. Such solution may not necessarily be better than the solution described in a), because the overall energy efficiency will be good only in a very narrow range of wind speeds.

For practical purposes, solution a) is then the most straightforward. Choosing an electric centrifugal pump requires just finding a suitable model from a catalog. Checking the operation curves provided by the manufacturer and selecting a model from these is the only action the engineer can take. The best pump for a given pumping total head and flow rate is the one where such design requirements intersect the operational curve as close as possible to the maximum efficiency point. For design purposes, the total head is the sum of the static lift (a.k.a. positive suction head), plus the static discharge head, plus the friction losses. Treating all details related to the selection of commercial centrifugal pumps is out of the scope of this book. The reader can find details on free online calculators and a free e-book to download in the Bibliography.

10.2.3 Positive Displacement Pumps

Positive displacement pumps can be of two different kinds: reciprocating (piston, plunger, bladder, and diaphragm pumps) and rotating (peristaltic, chain and washer a.k.a. rope pump, rotary vane, lobes, gears, and screw pumps in all their variants). Their flow rate is directly proportional to the number of strokes of the piston or the diaphragm or, in the case of peristaltic and rope pumps, to the rotational speed. Since water is incompressible, the pressure generated by such pumps can theoretically grow with no limits, risking damage to the pump's transmission organs or to the pipelines, a fact particularly true for piston pumps. In this section, we are going to analyze only piston, diaphragm, peristaltic and rope pumps, because these are the most common types employed with small wind turbines, and the easiest for self-construction.

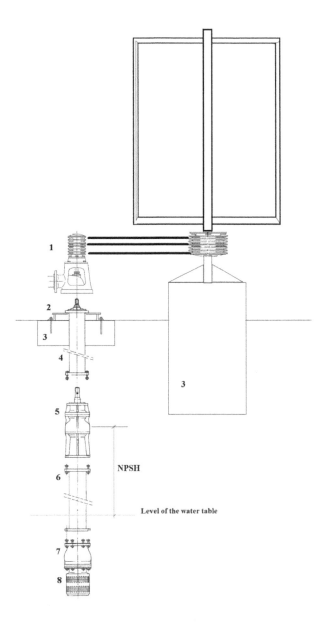

FIGURE 10.1
Shaft-driven submerged centrifugal pump with standard pulley and belt coupling designed for diesel motor or tractor's power take-off shaft. It is theoretically possible to replace the motor with a vertical axis wind turbine designed on purpose for such scope. Conceptual arrangement by the author; not to scale; borehole walls not shown. *Legend*: (1) pulley-shaft coupling and water discharge; (2) mounting flange with anchor rods; (3) concrete bases; (4) sections of shafted tube (necessary number to reach the water table); (5) pump rotor(s) (necessary number to overcome the foreseen water head, only one rotor shown here); (6) suction tube section; (7) bottom check valve; (8) suction screen; NPSH, Net positive suction head.

10.2.3.1 Piston Pumps

Piston pumps are of two kinds: single effect and double effect, being the first one the most diffused solution for pumping from deep boreholes. The operation principle of the piston pump is very simple: when the piston moves up, the displacement creates vacuum and the water pushes open the suction valve, entering into the cylinder. Simultaneously, the upward motion lifts the water contained in the riser tube. When the piston moves down, the suction valve closes, and the water exits the cylinder through the outlet valve, replenishing the riser tube. The use of these pumps is limited to clean water, since macroscopic suspended solids can clog the valves. The vacuum generated by the piston's movement is imperfect hence, although these pumps are self-priming, they cannot aspire water from depths bigger than 8 m. Pumps installed in boreholes usually have the piston submerged in water, in order to minimize the force required for the aspiration. As can be seen from Figure 10.2, mechanical work is necessary only during the ascending stroke, because the weight of the piston and connecting rods is enough for the descending stroke. Consequently, single effect piston pumps absorb power intermittently, and hence require some kind of compensation system in order to avoid pulsating accelerations of the wind turbine's rotor and fatigue effects on the mechanical parts. An additional problem of single effect piston pumps is the effect known as "water hammering" at certain speeds, caused by the pulsating acceleration and deceleration of the water column, especially in very deep wells. The basic calculation of the water hammer forces and the expedients to minimize damage risks will be dealt in Section 10.2.3.

Figure 10.3 shows a submerged double effect piston pump, less diffused than the simpler single effect pump that has been employed sometimes in Latin America, India, Australia, and Africa. Such kind of pump has the advantage of absorbing a more regular torque from the rotor, but requires that the connection rod works under compression during the down stroke, limiting the pumping depth to 20–30 m, otherwise the rod will buckle and eventually touch the well's walls, wearing out with time.

Another interesting solution is the one developed by the Colombian research center *Las Gaviotas* under the United Nations Development Program (drawings can be downloaded for free, see Bibliography). This pump must remain submerged in water at least 1 m, since the same enters from the bottom by natural pressure, through a standard 1″ foot valve with screen to prevent sand and grit entering the pump's chamber. It is a small diameter, double effect pump that employs ½″ PVC tubes as action rods, which at the same time lift water during the upstroke. The pump body is made of 1″ diameter stainless steel tube, joined to the PVC guide pipe of the same diameter, which lifts water during the downstroke. The piston-plunger head is made from a lathed Teflon bar. It can work up to 25 m depth. The small volume pumped in each stroke makes it suitable for direct coupling to the wind

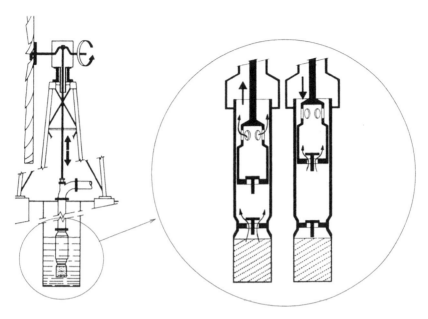

FIGURE 10.2
Conventional windmill with submerged single effect piston pump.

turbine rotor, increasing the overall efficiency. The *Gaviotas* windmill is able to start pumping from 6 m depth with winds as weak as 2 m/s.

10.2.3.2 Diaphragm Pumps

When it is necessary to lift water from shallow wells, canals, or rivers, diaphragm pumps are an interesting option. The operation principle is shown in Figure 10.4. Cheap commercial models of such pumps are available in the market, usually operated by means of handles or pedals, and can be adapted to any slow wind turbine. If necessary, it is also possible to make a diaphragm pump with locally available scrap materials and simple tools: old inner tubes from dismissed tires provide rubber of excellent quality for making the diaphragm and flap valves, while the body can be made of plywood, conveniently coated with polyurethane lacquer. As any other reciprocating pump, diaphragm pumps can be single action or double action. Double action diaphragm pumps are nothing but two single action units driven by means of a cantilever beam, Scottish yoke, or any other suitable reciprocating mechanism. The net positive suction head of these pumps is in the range 6–8 m. If built with generous inlet and outlet sections and flap valves, they can handle water with small solids in suspension without clogging. Since a diaphragm is like a piston with short stroke and big bore, it is easy to design it on purpose, in order to match the torque/speed curve of the wind turbine.

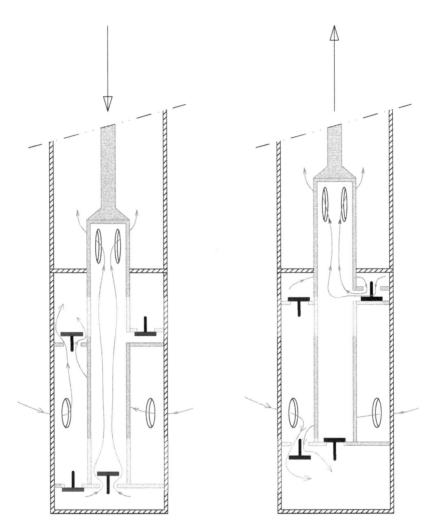

FIGURE 10.3
Operation principle of the double effect piston pump. The red section corresponds to the piston, which in this kind of pumps is always hollow to allow water passing through it and into the pipeline. There are several configurations of double effect pistons suitable for their installation in boreholes; some are very complex and patented. The one shown in this figure is purely conceptual; it sucks water from the sides.

In this way, a gearbox coupling is not necessary and the overall efficiency of the pumping system increases.

10.2.3.3 Peristaltic Pumps

The peristaltic pump's operational principle bases on the alternating compression and relaxation of a rubber tube, drawing the liquid in a similar way

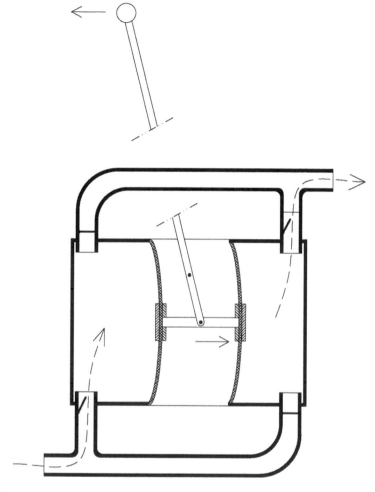

FIGURE 10.4
Sketch of the operation principle of an elementary double action diaphragm pump.

to our throat and intestines. A rotating shoe or roller passes along the length of the tube totally compressing it and creating a seal between the suction and discharge sides of the pump, eliminating product slip. Upon restitution of the tube, a strong vacuum is formed, drawing the liquid into the pump. The medium to be pumped does not come into contact with any moving parts and is totally contained within a robust, heavy-duty elastomeric tube. Such pumps have pressure ratings up to 16 bar (hose of elastomer material reinforced with a braid of nylon or similar fiber) and 2 bar (extruded rubber tube). Peristaltic pumps have no valves and hence do not clog, therefore, they can lift muddy water, and industrial models can handle materials as thick as mine slurry and wastewater sludge. Their lifting capacity ranges between

4 and 8 m, depending on the tube or hose's elasticity. Their efficiency with small suction lift and discharge heads up to 20 m can be as high as 88%. The flow rate is proportional to the rotor's speed and section of the tube or hose, the suction and discharge pressure are directly proportional to the thickness of the tube or hose, while the efficiency is inversely proportional to the same (more work to deform the elastomeric material for a given inner section of the tube). Because of their low rotating speed and high torque, as well as their constructive simplicity, peristaltic pumps are a suitable candidate for self-made *Savonius* pumps. Figure 10.5 illustrates a possible solution for self-construction.

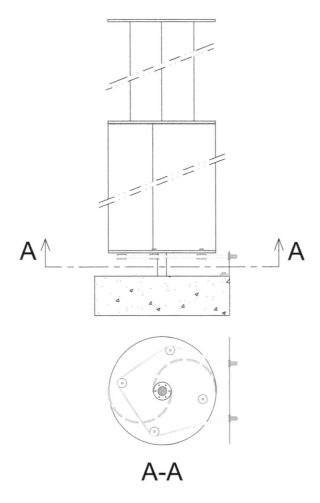

FIGURE 10.5
View from the bottom of a *Savonius* rotor section, showing how to employ the round plywood base, four pulleys bolted to it, a rubber tube, two garden hose connectors and an *L* steel profile fixed to the concrete base, to make a simple and cheap peristaltic pump.

10.2.3.4 Rope Pumps

Also called chain and washer pumps, or paternoster pumps, this type of rotary positive displacement pumps is much diffused in developing countries, because of its low cost and easy construction. The principle of operation is shown in Figure 10.6; the efficiency can reach 80%, the flow rate is proportional to the rotation speed and to the section of the tube. Too loose or too tight fitting of the washers into the tube results in lower efficiency, in the first case because a fraction of the water slips back, in the second because of the increased friction against the riser's wall. They are suitable for lifting

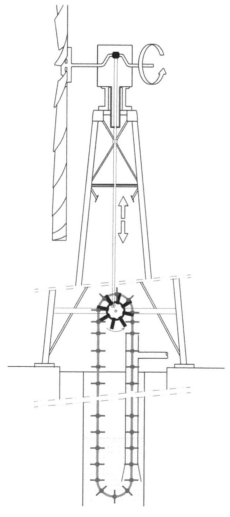

FIGURE 10.6
Conceptual view of a rope pump assembly.

1–100 m water heads, preferably on a vertical well. For low lifts, loose fitting washers are good enough to lift water efficiently through the pipe, since the back-flow will remain as a small and acceptable fraction of the total flow, while keeping friction to a minimum. At higher lifts, however, tighter fitting plugs rather than washers are necessary to minimize back leakage; many materials have been tried, but rubber or leather washers supported by smaller diameter metal discs are commonly used. Most chain and washer pumps have a bell mouth at the base of the riser pipe to guide the washers smoothly into the pipe. With higher lift units, a tighter fit is needed only near the lower end of the riser pipe; therefore, the riser pipe usually tapers to a larger diameter for the upper sections to minimize friction.

10.2.3.5 Bladder Pumps

Bladder pumps are a special kind of plunger pump where the transmission is hydraulic or pneumatic (Figure 10.7). A small piston is placed at the well's mouth, connected by means of a tube to a rubber ball or cylinder placed

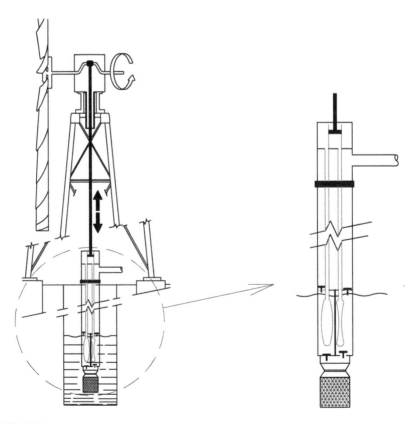

FIGURE 10.7
Double effect bladder pump.

inside a rigid cylinder with valves. The piston can pump either water or air to inflate and empty the bladder, but water is preferable because it is incompressible. Oil must be avoided to prevent the contamination of the water in case of accidental damage to the tube or bladder. Pedal-driven versions of such pumps have been employed in several rural aid projects in Africa, having proven to be robust and easy to service with local resources.

10.2.3.6 Bellow Pumps

Bellow pumps share some features of diaphragm pumps and piston pumps: they are made of rubber or similar flexible materials like the first ones, and feature long strokes like the latter. Commercial bellow pumps are usually meant for pumping small volumes of water, mostly are pedal-driven. Bigger bellow pumps have been employed in developing countries, according to a model designed by FAO's cooperation programs, shown in Figure 10.8. Two of such pumps connected with a cantilever beam constitute a double effect pump. Such pumps are suitable for low suction depths, low to moderate heads, and large pumped volume per single stroke. The efficiency depends on how stiff the bellow is: stiffer bellows require bigger force for the same displaced water volume of piston or diaphragm pumps. Soft bellows have efficiencies in the range 80%–85%, but their maximum suction and lift heads are usually lower compared to diaphragm pumps.

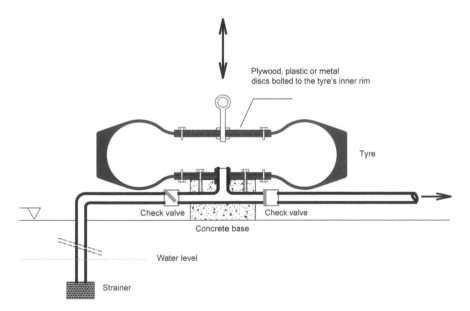

FIGURE 10.8
Section of a bellow pump built with an old car or motorcycle tire. Drawing by the author based on Fraenkel, P.L.; *FAO Irrigation and Drainage Paper 43, Water lifting Devices.* Such pumps require very high driving force and perform well with suction height <1 m.

10.3 Matching Hydraulic Pumps to Wind Turbines

When designing a wind pumping system with positive displacement pumps, it is fundamental to minimize the pump's starting torque, in order to allow the turbine to start rotating at low wind speed. The first step is to design the transmission organs that convert the turbine's rotation into a reciprocating motion suitable for the piston. As an example, consider a generic transmission system like the one depicted in Figure 10.9. Assume employing a single effect piston pump, since it is the most frequent case. The force on the piston in a single effect pump is not uniform: it is maximum during the upstroke—when the water is at the same time aspired to the cylinder and effectively lifted to the riser—while it is minimum during the downstroke, when the water is still inside the piston. The following procedure shows how

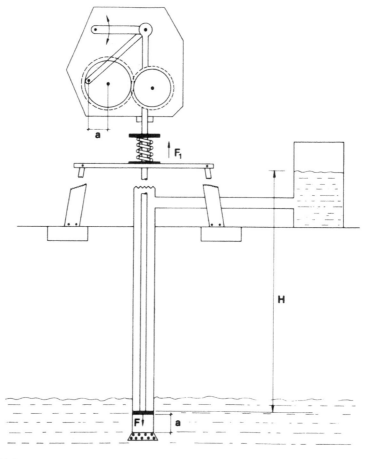

FIGURE 10.9
Using a spring to compensate the weight of the transmission rod, piston, and water column.

to calculate the maximum vertical force required from the turbine in order to overcome the weights of the transmission rod, of the piston, and of the water column:

$$F_{max} = W + \rho \cdot g \cdot S \cdot H$$

where:
F_{max} = vertical force required from the turbine (N)
W = weight of the transmission rod and piston (N)
ρ = density of water $\approx$ 1,000 kg/m³
g = gravity acceleration = 9.8 m/s²
S = section of the water column (m²)
H = piezometric head (m)

Being a the radius of the crank, the maximum torque is given by the following equation:

$$M_{max} = a \cdot F_{max} = a \cdot (P + \rho \cdot g \cdot S \cdot H)$$

Under such conditions, if d_1 is the diameter of the gear at the turbine's shaft, and d_2 is the diameter of gear coupled to the crank, the ratio between the rotation speed of the turbine's shaft, N_1, and of the crank, N_2, and consequently the ratio between the torque on each shaft, is given by the quotient:

$$k = \frac{d_1}{d_2} = \frac{N_1}{N_2} = \frac{M_2}{M_1}$$

The coefficient k allows us to calculate the torque M'_{max} that the turbine must produce in order to drive the pumping system. In general, the value of k can range from 1 (direct coupling) to 10 (gears with standard module equal to 3). Observe that M'_{max} is simply:

$$M'_{max} = \frac{a}{k} \cdot (W + \rho \cdot g \cdot S \cdot H) \tag{10.1}$$

In order to minimize the value of M'_{max} and allow the turbine to start with weak winds, we can choose among three different strategies, as shown below.

10.3.1 Minimizing a

If we reduce the radius of the crank, a, the piston's stroke will be shorter and consequently the volume pumped in each cycle will be smaller. The volume of the pump, called displacement, is given by the following formula:

$$q = 2 \cdot a \cdot S \tag{10.2}$$

where S is the section of the cylinder. The flow rate, Q, produced by a single action piston having section S, is just the product of the displacement by the number of piston strokes in the unit of time, N_2.

$$Q = q \cdot N_2 \rightarrow q = \frac{k \cdot Q}{N_1} \tag{10.3}$$

Hence, the section of the cylinder necessary to pump a given flow rate can be derived from Equations 10.2 and 10.3:

$$S = \frac{Q}{2 \cdot a \cdot N_2} = \frac{k \cdot Q}{2 \cdot a \cdot N_1} \tag{10.4}$$

Conclusion

By reducing a, we reduce the start torque demanded from the turbine, but this means increasing proportionally the piston's section in order to maintain the desired flow rate, consequently Equation 10.4 must be satisfied. In practice, if we need to pump from a shallow well or from superficial waters, increasing the section of the piston is not a problem. Eventually a diaphragm pump can be considered instead, since these behave like large diameter pistons with short strokes. If it is necessary to pump from a deep borehole, the cost of the drilling becomes more important, because this one is proportional to the borehole's diameter. The usual practice of wind pumps manufacturers involves offering two or three models of pistons with different strokes and displacements: short stroke and big diameter for small heads, long stroke and small diameter for pumping from deep wells.

10.3.2 Maximizing k

If we increase the multiplication factor, k, the number of piston strokes in the unit of time decreases, because the rotation speed of the crank-connection rod system is reduced proportionally.

From Equation 10.4 we can observe that the increase of k requires necessarily to increase the section S, in order to reach the desired flow rate Q.

Conclusion

Increasing the multiplication factor, k, has the same advantages and drawbacks of reducing the crank's radius already explained in Section 10.2.3.1. Usual gearboxes have $5 < k < 10$. Nevertheless, keeping low values of k has the advantage that the diameter of the gears, and consequently the weight and cost of the transmission system, will be lower. Ideally, for $k = 1$ there is no need to employ gears, so it is possible to save their cost, but the pump will wear out faster because of the higher frequency of the piston's strokes. This is not a problem as the pump is self-built.

10.3.3 Regulation of the Torque between the Extreme Values M_{min} and M_{max}

Since the single effect piston pump is the simplest and cheapest, most manufacturers have adopted it, but the issue to be solved is that the pump absorbs energy from the turbine only during the piston's upstroke. This has a repercussion on the start torque too: if the rotor is static and the piston is at the lowest point of its stroke, the necessary torque at the shaft to start the rotation will be M_{max}, and in the opposite case, it will be M_{min}.

The possible solutions to obtain a uniform torque are:

a. employing a double action pump, but this one is subject to the limitations already described in paragraph Section 10.2.1, or;

b. installing a spring or, less frequently, a counterweight.

Figure 10.9 shows the most common system, using a precompressed spring in order to counteract the total force F. In this way, the motor torque will be more regular, and the start torque will be independent of the piston's initial position.

The operation principle of the spring compensator is based on the equations already introduced in Section 10.2.3, modified with the addition of the spring's force, F_1. The force applied by the connecting rod during the piston's upstroke, and the consequent maximum torque at the crank's shaft, are the following:

$$F = W + \rho \cdot g \cdot S \cdot H - F_1 \quad \text{and} \quad M_{up} = a \cdot F = a \cdot (W + \rho \cdot g \cdot S \cdot H - F_1)$$

During the piston's downstroke, the spring is still acting, but produces a resisting effect, F_2, higher than the total weight of the connecting rod, piston, and water column. Therefore, during the piston's downstroke the torque at the shaft is resisting too, and its value is:

$$M_{down} = a \cdot F_{down} = a \cdot (W - F_2)$$

Observe that the term $\rho \cdot g \cdot S \cdot H$ is null, because during the downstroke of a single effect pump, the water simply flows inside the piston and the friction resistance of the open valve is negligible.

The wind turbine's rotary motion will be most regular when the torque at its shaft during the upstroke will be equal to the torque during the downstroke of the piston, i.e.,

$$M'_{up} = M'_{down} \rightarrow a \cdot (W - F_2) = a \cdot (W + \rho \cdot g \cdot S \cdot H - F_1)$$

Simplifying both members, we obtain:

$$F_1 - F_2 = \rho \cdot g \cdot S \cdot H$$

If the element for the compensation of the forces is a steel spring, its force is given by Hooke's Law, so the following equation must be satisfied:

$$K \cdot (x_1 - x_2) = \rho \cdot g \cdot S \cdot H$$

where:

K = elastic constant of the spring (N/m)

x_1 = initial deformation of the spring at the stroke's highest point (m)

$x_2 = x_1 + 2 \cdot a$ = final deformation of the spring at the lowest point of the stroke (m)

It is easy to demonstrate that, at the highest point of the piston's stroke, the following condition is satisfied:

$$K \cdot x_1 = W$$

And the following equation is valid at the lowest point of the piston's course:

$$K \cdot (x_1 - x_1 - 2 \cdot a) = \rho \cdot g \cdot S \cdot H$$

With both border conditions, it is now possible to determine the necessary values of K and x_1 to design the spring.

If the designer prefers to adopt a compensation system with counterweight, a solution usually employed for very deep wells, it is easy to demonstrate that its weight, W_c, must be equal to:

$$W_c = W + \frac{\rho \cdot g \cdot S \cdot H}{2}$$

10.3.4 Systems for the Conversion and Transmission of the Motion

Once the values of a, k, and S, are defined, and the compensation system to regulate the torque is calculated, it is only necessary to decide how to transform the rotary movement of the turbine in a suitable reciprocating movement of the piston.

The possible alternatives are: employing a crank-rod system, a camshaft, or a Scotch yoke. The first is the simplest to build, but tends to displace the rod laterally as well as vertically. In order to avoid—or at least to minimize—such undesired effect, it is possible to resort to lever or guide systems, as the ones depicted at the left and center of Figure 10.10.

The transmission systems depicted in Figure 10.10 convert a rotating motion into a reciprocating motion, but both the linear speed and corresponding driving force vary according to a sinusoidal law. The system shown in Figure 10.11, called rack and pinion mechanism, overcomes such limitation: if the rotation speed is constant, then the translation speed is

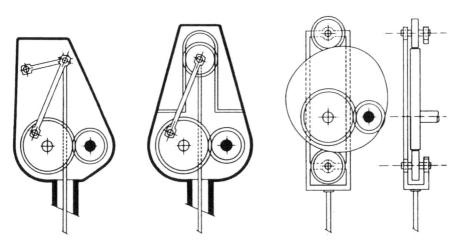

FIGURE 10.10
Different mechanisms are able to keep a straight motion of the vertical transmission rod.

constant both during upstroke and downstroke, hence the force's magnitude is constant, and its variation in time corresponds to a square wave: half of the cycle positive (up) and half negative (down). The energy of a square wave is higher than that of a sinusoidal wave of the same amplitude and period, so the transmission in Figure 10.11 is potentially more efficient than the ones in Figure 10.10. Its drawbacks are a higher manufacturing complexity, and being more prone to the phenomenon known as water hammer (see Section 10.2.3.5).

Some wind pumps proposed by Latin American and Australian manufacturers, employ hydraulic instead of mechanical transmission. The working fluid in such hydraulic systems is water instead of oil, in order to avoid the risk of contamination of the well in case of accidental damage of the transmission circuit. Figure 10.12 shows schematically the working principle of a solution proposed in Colombia by the NGO *Ingenieros sin Fronteras (Borderless Engineers)*. Observe that the spring shown in Figure 10.9, in this case has been replaced by a gas spring, i.e., a closed volume where air is alternatingly compressed and expanded. In the Author's opinion, such solution is worse than employing a steel spring, because the process of compression and expansion of gases is thermodynamically irreversible. Hence, the alternating compression and expansion of the air in the compensation chamber dissipates energy as heat that can even stop the rotation of the turbine shaft because of the overpressure on the piston's head caused by the overheated air. The bladder pump, already explained in Section 10.3.5, is a better solution when employing hydraulic transmissions.

Some authors have proposed the use of a pneumatic instead of a hydraulic transmission. The operation principle is the same in both systems, with a slight difference in the submerged cylinder, but the pneumatic transmission

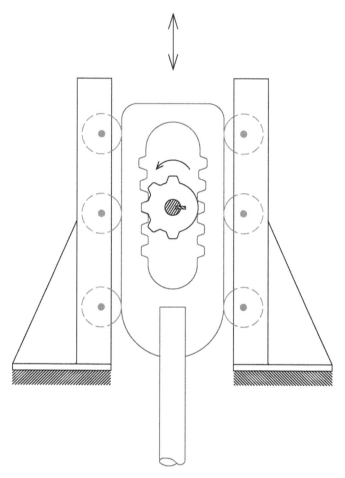

FIGURE 10.11
Reciprocating rack and pinion mechanism.

is less efficient because of the heat dissipated in the thermodynamic irreversibility of the process. Its advantage is a lighter construction and no water hammering, because the compressibility of air absorbs the shocks caused by the water hammer effect.

10.3.4.1 Sizing the Transmission between the Wind Turbine and the Pump

In order to size exactly the transmission organs between turbine and pump, it is necessary to know the rotation speed of the first as a function of the wind speed, and the flow rate of the second as a function of the rotation speed of its crank. From the statistic distribution curves of the wind speed, considering only those bigger than the turbine's cut-in speed, it is possible to determine the volume of pumped water by means of discrete integration

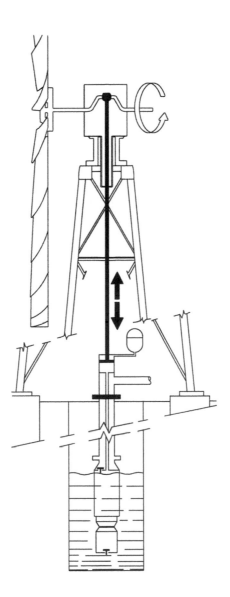

FIGURE 10.12
Working principle of the single effect hydraulic transmission with compensation by air spring, developed in Colombia by the NGO *Ingenieros Sin Fronteras*. Graphic re-elaboration by the author.

(for instance, with a spreadsheet) in a given period of time (one month, one year, etc.).

10.3.4.1.1 Determination of the Cut-In Wind Speed

The input datum is the maximum torque when the rotor is static, M_{max}, dependent on the chosen mechanism (see Section 10.3.3).

Knowing also the curve of C_P as a function of λ, and because of the relationships demonstrated in Section 2.4.3.6, we will have the following formula:

$$C_P = C_M \cdot \lambda$$

Therefore, it turns relatively easy to plot the driving torque, M (Nm), as a function of N (R.P.M) for different values of wind speed (assumed as constant) with the help of a spreadsheet. We will obtain curves like the one shown in Figure 10.13.

Figure 10.14 shows the graphic procedure. Draw a horizontal line passing through M_{max}, calculated according to the formulas explained at the

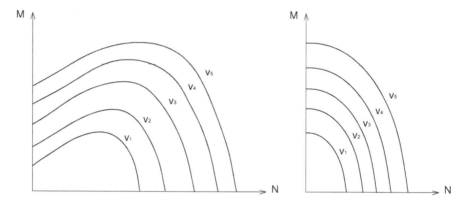

FIGURE 10.13
Characteristic shapes of the curves of M as a function of N, using the wind speed V as parameter, for an American windmill (left) and a *Savonius* rotor (right).

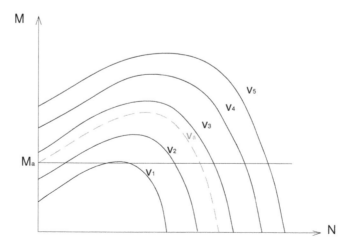

FIGURE 10.14
Graphic procedure to calculate the cut-in speed.

beginning of Section 10.3. Such value must coincide with the starting torque, M_a. Now search the curve of wind speed that intersects the point M_a. Such wind speed will be the cut-in speed of the system, V_a (usually called V_{cut-in} in the catalogs of wind turbine manufacturers). Winds weaker than V_a will not be enough to start the rotation of the turbine.

In general, it is improbable that the horizontal line with origin in M_a intersects a curve corresponding to an integer value of wind speed. In order to interpolate between the closest curves and to determine the exact value of V_a, the definition of C_M must be employed:

$$C_M = \frac{2 \cdot M}{\rho \cdot \pi \cdot R^3 \cdot V^2}$$

Since C_M is a constant value (independent of the wind speed) for $\lambda = 0$ (stationary turbine), we can replace M with M_a in the former formula and then calculate the speed V, which corresponds to the desired value V_a.

Now that the cut-in speed V_a is known, it is possible to plot the curve of M as a function of N, characteristic of the turbine-transmission-pump system (dotted line in Figure 10.13).

10.3.4.1.2 Determination of the Rotary Speed under Normal Operation

The rotary speed under normal operation can be plotted with the help of the torque versus speed curve obtained in the former Section 10.3.4.1.1, or by means of a power versus rotary speed diagram, taking the wind speed as parameter.

a. Torque versus speed curve

Determine the average resistant torque, $\bar{M}_r$, on the turbine's shaft, caused by the pump and the transmission system under normal operation conditions, using the following formula:

$$\bar{M}_r = \frac{P}{\Omega} = \frac{\rho \cdot g \cdot Q \cdot H}{2 \cdot \pi \cdot \frac{N}{60} \cdot \eta}$$

where:
   ρ = density of water $\approx 1,000\,\text{kg/m}^3$
   g = gravity acceleration = $9.8\,\text{m/s}^2$
   Q = flow rate (m³/s)
   H = total pumping head (piezometric + losses in the pipelines) (m)
   N = rotary speed (R.P.M)
   60 = s/min
   η = mechanical efficiency (pump plus transmission) $\approx 90\%$ for piston pumps
   P = power at the shaft (W)
   Ω = angular speed (rad/s)

Plotting on the diagram the horizontal line with ordinate $\bar{M}_r$ (Figure 10.15) we determine the intersections between the said line and the curves V_a, V_3, V_4, etc., which correspond to the values of the rotary speeds N_a, N_3, N_4, ... N_n, measured on the abscise axis.

b. Curve of the power as a function of the rotary speed

The family of the curves of power as a function of the rotary speed can be obtained from the curve of C_p as a function of λ, taking as individual parameter of each curve of the wind speed. The power absorbed by piston pumps varies with the speed of rotation of the crank, according to the following expression:

$$P = \Omega \cdot \bar{M}_r = \frac{\rho \cdot g \cdot H \cdot k \cdot q \cdot N}{\eta \cdot 60}$$

where:

ρ = density of water $\approx 1{,}000 \, \text{kg/m}^3$
g = acceleration of gravity = $9.8 \, \text{m/s}^2$
H = total pumping head (piezometric + losses in the pipelines) (m)
k = multiplication ratio of the transmission between the pump and the turbine
q = displacement of the pump (m³)
N = rotary speed (R.P.M)
60 = s/min
η = mechanical efficiency (pump plus transmission) $\approx 90\%$ for piston pumps

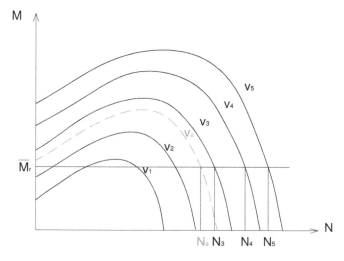

FIGURE 10.15
Graphic calculation procedure of the rotary speed of the turbine for each value of the wind speed, using the curves of *M* as a function of *N*.

Observe that the expression of P as a function of N corresponds to a straight line passing through the origin of the Cartesian axes system, whose slope is proportional to k. Plotting the said straight line, the interceptions with the power curves at the points $P_1, P_2, P_3...$ correspond to the rotary speeds $N_1, N_2, N_3...$ (Figure 10.16).

10.3.4.1.3 Determination of the Optimum Reduction Ratio, k

Most slow wind turbines turn at a speed higher than the one necessary to drive a piston pump. In the former sections, we have assumed that the factor of speed reduction, k, was known. In reality, if the factor k is not fixed by the available mechanism (e.g., a standard pump with built-in gearbox), it is convenient to calculate it for the maximum efficiency under any wind speed, and then build the pump accordingly.

The procedure to optimize k is the following:

a. Determine the power versus rotary speed family of curves, and draw the straight line of the pump's power. The equation of the said straight line is:

$$P = \Omega \cdot \bar{M}_r = \frac{\rho \cdot g \cdot H \cdot q \cdot N}{\eta \cdot 60}$$

Such line corresponds to $k = 1$, and represents the mechanical power output of the turbine running at N turns per minute, as if it was directly coupled to the pump.

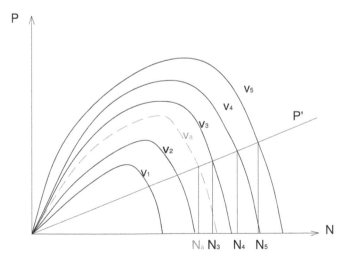

FIGURE 10.16
Determination of the pumping power and the rotary speed at different wind speeds.

In general, the said straight line will never intersect the maximum points of the P versus N curves. In the example shown in Figure 10.16 we can observe that, if the rotor were directly coupled to the pump, the useful power would be too low and the flow rate effectively pumped would be insignificant, because the turbine is unsuitable to dive the pump in question.

When coupling the pump by means of any multiplication mechanism, the ratio between the speed N_1 (rotor) and N_2 (pump) is the following:

$$N_1 = k \cdot N_2$$

At this point, we need to consider the unavoidable friction loss in the transmission system; hence, the power that the rotor must provide to the pump will be:

$$P' = \frac{P}{\eta'} = \frac{\rho \cdot g \cdot H \cdot q \cdot k \cdot N_2}{\eta' \cdot \eta \cdot 60}$$

where:
η' = efficiency of the transmission

We must choose such a value of k that the straight line P' will pass as close as possible to the maximum points of the power curves (Figure 10.17).

The average rotary speed of the wind turbine is represented by the values measured on the abscise axis corresponding to the intersection points of the straight line P' and the P vs. N characteristic curves.

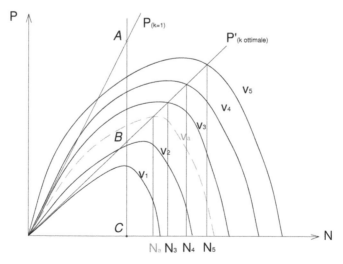

FIGURE 10.17
Graphic procedure to determine the optimum reduction factor k.

In order to calculate k, take any point A along the line P, draw a vertical line until intersecting the line P' in the point B and the abscise axis in the point C. The value of k is given by the quotient:

$$k = \frac{\overline{BC}}{\overline{AC}}$$

Knowing the annual (or monthly) duration (in hours) of the winds having speeds $V_1, V_2, V_3...$, the speeds $N_1, N_2, N_3...$ are univocally to them, and hence it is possible to calculate immediately the annual (or monthly) flow rate that the system will be able to pump.

10.3.5 Minimizing Water Hammering in the Pipeline and Check Valves

Water hammering is a second-order problem in usual wind pumping systems, because the diameters and lengths of the tubes are small enough to exclude the possibility of pressure surges able to damage the system. Nevertheless, it cannot be neglected, because pressure waves travelling along the pipelines reduce the overall efficiency of the pumping system. Dealing in depth with the theory of water hammer is out of the scope of this book. The reader can find a simplified explanation in Fraenkel, P.L.; *FAO Irrigation and Drainage Paper 43, Water lifting Devices*, (Section 3.5.1) and Lüdecke, H.J., Kothe, B.; *Water Hammer*, KSB AG Communications, Halle, Germany, 2015 (see Bibliography and links for free downloading). For practical purposes, when designing a pumping system with single effect piston pumps it is necessary to take into account the following notions:

a. Single effect piston pumps create a pulsating flow. The water inside the tube is subject to an acceleration (upwards in the riser, forward in horizontal tubes) during the upstroke and then to an acceleration in the opposite sense during the downstroke.

b. Such accelerations induce pressure waves that travel along the tubes. Under certain conditions, the peak pressure can be several times the static head, which in deep wells can damage the pump's check valves. The valley pressure can be low enough to induce cavitation, which can damage the piston's head and tube's walls.

c. Such waves dissipate energy as additional friction against the pipeline walls, increasing the effective head and reducing the flow rate.

d. The overpressure is proportional to the square of the piston's frequency (hence to the square of the turbine's rotary speed) and to the total length of the tubes.

e. The pressure variation can be calculated with Joukowsky's equation:

$$\Delta P = \rho \cdot c \cdot \Delta v$$

where:

    ΔP = pressure variation (Pa)

    ρ = density of water = 1,000 kg/m³

    c = speed of sound in water (m/s)

    Δv = variation of the flow's speed (m/s)

The speed of sound in unconfined water is 1,440 m/s at 10°C. When it is confined in steel tubes, its value is nearly 1,400 m/s. If the tubes are made of plastic, or if one end of the tube is in contact with the atmosphere (or an air chamber), it sinks to just 300 m/s. The amount of dissolved air (or other gases) in the water also plays an important role. In water containing just 0.2% of dissolved gases, the speed of sound drops to 470 m/s. As a rule of thumb for quick evaluation purposes, the value of a in steel tubes and clean water is usually rounded to 1,000 m/s.

f. Joukowsky's equation is valid only when the time t (in seconds), in which the fluid is changing speed, is shorter than the time required for the shock waves to travel the full length of the tubes, L. In mathematical form, such condition is expressed as:

$$t < \frac{L}{c}$$

10.3.5.1 Numerical Example

A windmill is pumping from a 50 m-deep borehole to a tank placed 10 m above ground. The piston's stroke is 30 cm. The speed of the wind is such that the piston's frequency is 2 strokes/s. Calculate the overpressure on the piston's head and check the valve.

The first step is to check if the length of the tube allows water hammering at all. Since the piston of a wind pump is connected to a crank, its displacement speed follows a sinusoidal law with ½ s period. This means that the piston will take 0.5 s to travel from one end to the other, starting and finishing with null speed. The maximum speed is then at the middle point of the stroke, so the resulting time, t, for the maximum variation of v is 0.25 s. Comparing t with the time that a shock wave would take to travel along the pipeline, we obtain:

$$\frac{60 \text{ m}}{1,000 \text{ m/s}} = 0.06 \text{ s} < 0.25 \text{ s}$$

Hence, the Joukowsky equation is not valid. There will be indeed some overpressure at the extreme positions of the stroke, but it will be very small, in the order of a few meters of water column. Should the wind become so strong that the rotation speed of the crankshaft exceeds 16 turns/s (960 R.P.M.), then the pump and pipeline would be subject to water hammering according to the Joukowsky equation. When determining the speed reduction factor k, it is then necessary to check that the rotation speed at the pump's crankshaft does not exceed such value, which is unlikely in most practical cases.

10.4 Examples of Design of Wind-Powered Pumping Systems

The designer of a wind-powered water pumping system cannot overlook the definition of the most suitable technology for each particular case. Tables 10.1 and 10.2 are a useful guide for choosing the configuration of the system.

Observe that, from the point of view of the energy efficiency, a modern fast turbine with electric generator driving an electric centrifugal pump through inverter, and a traditional American-type windmill or even a Cretan sail rotor with a homemade piston pump, are equivalent. A slow horizontal axis wind turbine designed on purpose to drive directly a piston pump has higher efficiency than current electric wind-powered pumping stations, because there will be less conversions of energy between the turbine's shaft and the delivered water volume at the desired height. The final decision on which system to be adopted requires considering other non-technical factors, as shown in Table 10.3.

Apart from defining *how* to pump the water, it is necessary to know *how much* water is required. Table 10.4 provides some guidelines for a reasonable estimation when the exact water consumption is unknown.

10.4.1 Example: Design of a Wind-Driven Pumping System in an Isolated Area

A dairy farm in Botswana has 120 cows. Some cows are productive, others are calves, heifers, or dry cows; hence, considering an average consumption of 80 l water/day is a realistic estimation. Therefore, the pumping system must deliver at least 9.6 m³/day. The borehole is 50 m deep, the pre-chosen solution is a piston pump driven by an American-type windmill, both because of cost questions and ease of maintenance and reparation on site.

10.4.1.1 Determination of the Average Pumping Power

$$P = \frac{Q \cdot \rho \cdot g \cdot H}{\eta}$$

$Q = 9.6$ m³/day $= 0.111$ /s
$\rho = 1,000$ kg/m³ $= 1$ kg/l
$g = 9.8$ m/s²

TABLE 10.1

Summary of the Usual Efficiencies of Different Turbines, Pumps, and Conversion Devices

Turbine	C_P
American-type windmill	0.3
Cretan windmill	0.3
Savonius	0.3
3-Blade horizontal axis fast turbine	0.42
Modern multi-blade horizontal axis slow turbine	0.38
3-Blade Darrieus	0.35

Pump	Typical Efficiency
Piston (single and double effect, Teflon piston rings on smooth plastic or steel)	0.9
Piston (homemade, rubber or leather piston rings on steel)	0.85
Diaphragm/bladder/soft bellow	0.8
Peristaltic, extruded rubber tube	0.8
Peristaltic (thick tube or hose)/hard bellow made with a vehicle tire	0.6
Rope and discs	0.8
Centrifugal (sub-optimum speed)	0.65
Centrifugal (at optimum speed)	0.8

Conversion/Transmission Device	Typical Efficiency
Crank/camshaft	0.98
Gearbox/pulley and belt (one step)	0.95
Hydraulic transmission	0.95
Pneumatic transmission ($P < 4$ bar)	0.87
Pneumatic transmission (4 bar $< P < 8$ bar)	0.84
Alternator <20 kW, permanent magnets, 24 poles, 3 phases, at full power	0.93
Inverter (at full power)	0.97
Squirrel cage motor. IEC efficiency class 2. $2 < P < 20$ kW (at full power)	0.8
Stationary lead-acid battery	0.7

$H = 50$ m (+10% of friction losses in the pipes)
$\eta = 90\%$ (good quality piston pump)

The average power is then 66.5 W and the total energy at the crankshaft is 1.6 kWh/day.

10.4.1.2 Sizing the Turbine

From the available meteorological information, the annual energy is 323 kWh/year·m². Since the C_P of American-type windmills is 0.25, the energy density at the turbine's shaft will be:

$$E_{\text{available}} = 0.25 \cdot 323 (\text{kWh/year} \cdot \text{m}^2) = 80.75 (\text{kWh/year} \cdot \text{m}^2)$$

TABLE 10.2

Comparison of the Overall Efficiency of Different Configurations of Wind Powered Pumping Systems

Configuration of the System	Global Efficiency (%)
Traditional American windmill + gear + piston pump	24
Savonius + rope, or peristaltic, or diaphragm pump, direct drive	24
Cretan + piston pump, direct crankshaft drive	25
Darrieus + pulley drive + centrifugal pump	22
Modern multi-blade turbine + piston pump, direct crankshaft drive	32
3-Blade horizontal axis wind turbines (HAWT) + alternator + battery + inverter + electric centrifugal pump	17
3-Blade HAWT + alternator + inverter + electric centrifugal pump (battery completely charged or no battery)	24

TABLE 10.3

Non-technical Factors to Consider when Choosing the Configuration of the Wind Powered Pumping System

Configuration	Cost	Existing Pump and Borehole	Weak Winds	Low Flowrate	Deep Borehole	Low Suction Head	Electric Grid Not Available
Slow HAWT + piston pump	Low–medium		x	x	x		
Savonius + diaphragm or rope, or peristaltic pump	Low		x			x	
3-Blade Darrieus, directly connected to submerged centrifugal pump	Medium –high	x				x	
Electric 3-blade wind turbine + centrifugal electric pump	Medium –high	x			x		x

The necessary energy for pumping the required quantity of water is:

$$E_{pump} = 1.6(kWh/day) \cdot 365(days) = 584(kWh/year)$$

Calculating the minimum area of the turbine, and its diameter, are straightforward:

$$S = \frac{584}{80.75} = 7.21\,m^2 \rightarrow D = 3.03\,m$$

TABLE 10.4

Indicative Daily Water Consumption for Farming and Residential Scopes

Scope of Pumping System	Water Consumption
Residential (minimum vital)	20 l/person·day
Residential (European urban standard, 1 shower/day, water saving appliances)	50 l/person·day
Residential (Urban standard including gardening, or bathing in tub, or several showers/day)	100 l/person·day
Dairy cattle, intensive farm, $t > 25°C$	150 l/animal·day
Cattle, extensive farm, $t < 25°C$	50 l/animal·day
Boar	15 l/animal·day
Pig (50–100 kg)	10 l/animal·day
Pregnant sow	20 l/animal·day
Sow with piglets	30 l/animal·day
Sheep/goat	10 l/animal·day
Horse	50 l/animal·day
Hen/chicken	0.25 l/animal·day
Turkey	0.5 l/animal·day
Rabbit (pregnant female, or female with cubs, dry diet)	2 l/animal·day
Rabbit (vegetable diet)	0.3 l/animal·day
Green grass, garden flowers (summer)	5 l/m²·day
Fruit trees	30 l/tree, once every 2 days
Tomatoes, lettuce, spinach (summer)	3–5 l/m²·day
Corn	3.5–5 l/m²·day
Sugar beet (spring/summer)	1.5/2.5 l/m²·day
Wine grapes	6.6 l/m²·day

Please note that the energy density of the site, obtained from meteorological sources, includes the energy in the interval 0–3 m/s, when the turbine is not producing. Considering such limitation, a commercial turbine (or a self-constructed Cretan windmill) having at least 4 m diameter should be adequate as preliminary sizing.

10.4.1.3 Determination of the Storage Volume

In order to calculate the volume of the storage tank, we can choose between two different approaches:

a. Analyze the seasonal variations of wind speed and define a storage volume equal to the sum of all deficit months. In this case, it is necessary to know at least the monthly average wind speed. If the tank is open, it will collect precipitations but also lose water by evaporation. So, it is necessary to know the daily temperature, pressure, relative moisture, and rainfall.

b. If daily meteorological data of a typical year are available for the site, it will be easier and more reliable to simulate with a spreadsheet for the flow rate and total volume pumped each day. Then check different storage volumes (supposing the tank is empty, or has an arbitrary initial volume) until the total volume stored is bigger than zero (or bigger than an arbitrary reserve level) all the time along the year. Alternatively, a bigger turbine with a smaller storage can turn to be more cost effective.

Furthermore, it is necessary to check that at the end of the typical year, the remaining stored volume is bigger than, or equal to, the initial volume. Otherwise, either increase the diameter of the turbine or the size of the storage tank, or evaluate the possibility of rain harvesting to integrate the pumped water.

When economically feasible, *flexytanks*, underground cisterns, or any other kind of closed tanks are the best option, both to eliminate evaporation and to maintain a good quality of water. If the budget is limited , or in the case that an "Australian tank," or an artificial pond are the foreseen or already existing options, then it is necessary to evaluate the evaporation losses. The additional volume that the turbine must pump in order to compensate the evaporation losses, can be calculated with the following empirical formula:

$$Q_{evap} = \Theta \cdot A \cdot (x_s - x)$$

where:
Q_{evap} = water evaporation flow rate (kg/h)
Θ = evaporation coefficient (kg/m²·h) resulting from the following empirical formula:

$$\Theta = (25 + 19 \cdot V)$$

V = wind speed at the water surface (m/s) (assume 10% of average wind speed)
A = area of the pool (m²)
x_s = absolute moisture of the saturated air at the temperature of the pool's surface (kg H_2O/kg dry air)
x = absolute moisture of air (kg H_2O/kg dry air) (consider average temperature and moisture of the site in the reference period of time)

In our example, suppose that the foreseen storage will be an Australian tank with 20 m diameter. During the dry season, the average temperature will be 32°C and the wind speed at the tank's surface will be 0.2 m/s. The temperature of the water surface can be estimated as, at least 25°C. The average relative moisture of the air during the dry season, according to the available meteorological data, is 35%. From the psychrometric chart, we find out that $x_s = 0.020\,\mathrm{kg/kg}$ dry air and $x = 0.011\,\mathrm{kg/kg}$ dry air. The calculation is straightforward:

$$A = \pi \cdot \frac{25^2}{4} = 491\ \mathrm{m}^2$$

$$\Theta = (25 + 19 \cdot 0.2) = 28.2\,(\mathrm{kg/m}^2 \cdot \mathrm{h})$$

$$Q_{evap} = 28.2 \cdot 491 \cdot (0.020 - 0.011) = 125\ \mathrm{kg/h} = 2{,}991 \cdot \mathrm{kg/day}$$

Conclusion

Employing an open pool for water storage, the most diffused practice in farming, will require pumping 32% more water than the demand assumed for design, just in order to compensate the evaporation. This means either employing a turbine with nearly 16% larger diameter, or considering the purchase of a tarpaulin cover for the tank to avoid evaporation losses. The second option should be preferred for four reasons:

a. The cover avoids dust, insects, and other animals polluting the water.

b. The cover will prevent the sun from heating the concrete bottom and hence the water, keeping the latter at a lower temperature. Animals drink less water when the same is fresh. Keeping the animals fresh relieves in part the stress that all bovines suffer with high temperatures, increasing the dairy's productivity.

c. Keeping the water in the dark avoids the proliferation of microscopic algae and cyanobacteria, which can give a bad taste to the water or in some extreme cases, make it toxic to animals and people.

d. Pumping just the necessary amount of water means a longer life of the borehole and higher resilience of the farm under eventual severe draughts, because overexploiting unnecessarily the aquifer can lead to drying it out.

10.4.2 Example of the Design of a Wind Pumping System for Industrial Agriculture

A farmer cultivates tomatoes in greenhouses for industrial customers. It is necessary to irrigate 4 ha of cultivation placed in Caltagirone, Italy. The water demand, according to the local experience, is 8 mm/day (August), 4 mm/day during the other months. The water table is 50 m deep and there is already a borehole equipped with a submerged centrifugal pump driven by a vertical shaft and a diesel motor. The consumption of diesel oil is 11,830 l/year. The farmer desires to reduce the cost of pumping by replacing the diesel fuel with wind power.

The possible solutions in such context are:

a. Replacing the diesel motor with an electrical one, and installing a wind turbine in parallel with the national grid. Current regulations in Italy allow to exchange energy along the year, so the excess of energy generated in winter, when the wind is stronger, can be balanced with the energy consumed from the grid in summer.

b. Coupling the shaft of a *Darrieus* turbine to the pump's shaft, and storing water in a tank.

In this case, solution a) is the most convenient, because the electricity supply in Italy is reliable. An inverter for paralleling a horizontal axis wind turbine with the grid is cheaper than a stand-alone inverter with static battery bank. Large Darrieus turbines are expensive and the need to couple it mechanically to an existing shaft requires a custom-made unit.

The reader will find attached to this chapter the file *Caltagirone-2009.xlsx*. The same contains the meteorological data and the calculations explained in the next sections.

10.4.2.1 Determination of the Average Power for Pumping

The first step is to calculate the water demand in August:

$$Q = 0.008\,(\text{m}/\text{day}) \cdot 4\,(\text{ha}) \cdot 10.000\,(\text{m}^2/\text{ha}) = 320\,(\text{m}^3/\text{day}) = 3.7\,(\text{l}/\text{s})$$

And consequently, $Q = 1.85$ (l/s) the rest of the year.

Since the pump is already installed, we can get its operational curves from the manufacturer's catalog. The pump is designed for the operation with a diesel motor, hence, the farmer used to adjust the rotation speed until obtaining the desired flow rate. If we now replace the diesel motor with an electrical one, there are two alternatives: purchasing a motor with inverter to adjust its speed, or a cheaper unit with no control. For budgetary reasons, the second option will be adopted. Since the frequency in Italy is 50 Hz, the maximum speed of a standard squirrel cage motor is 2,900 R.P.M. According to the pump's datasheet, with the rotor at 2,900 R.P.M. the pump can lift 4.2 l/s to 55 m, and the mechanical power at the shaft is 4.6 kW. Under such operational conditions, the mechanical efficiency is 49% and the electrical power necessary to pump 3.7 l/s can be calculated with the usual formula:

$$P = \frac{Q \cdot \rho \cdot g \cdot H}{\eta}$$

We assume $H = 55$ m, to account for the friction losses in the pipelines, and $\eta = 49\% \cdot 88\% = 43\%$ to account for the overall efficiency of the system, which is the product of the pump's mechanical efficiency by the electric efficiency of the motor. The absorbed electrical power is then:

$$P_{el} = 4.5(kW) \rightarrow E_{el} = 108(kWh/day) \quad \text{during August and}$$

$$P_{el} = 2.25(kW) \rightarrow E_{el} = 54(kWh/day) \quad \text{the rest of the year.}$$

Hence, the total energy demand for water pumping is 21,384 kWh/year (see file *Caltagirone.xlsx*, sheet *electricity production*).

10.4.2.2 Size of the Turbine

As a first approach, the balance between the electricity produced along one year by the turbine, and the electricity consumed by the pump, must be as close to null as possible. Depending on the feed-in tariff, in some cases there may be some convenience in producing more energy than the one consumed, while in other cases it may be the contrary. The optimum size of the installation will depend on the cost of the turbine, support structure, and inverter, and not only on the feed-in tariff. In order to perform a correct financial analysis, it is necessary to calculate the wind energy productivity of the site.

The available meteorological data are: the daily averages of wind speed, temperature, and pressure, corresponding to a whole year (2009). The raw data, as downloaded from the source, are presented in the file *Caltagirone. xlsx*, sheet *raw daily avg 2009*. The first step is to "clean" the data; in this case, converting all data in different units into coherent SI units, then replacing the wrong pressure values (null) by the standard pressure, 101,300 Pa. Now we have a table with coherent data, we can add a column where to calculate the average air density of each day. The result can be seen in the sheet *clean avg 2009*. The average daily wind speed is not suitable to calculate the available energy. For this reason, we need to calculate the average energy (or the average cubic wind speed) from the Weibull distribution, in which $N = 24\,h$ and V is the corresponding to each day in the year's table. In the sheet *clean avg 2009*, we will determine then the maximum and minimum value of the daily average V throughout the year, which turns to be 7.2 and 1.1, respectively. Now, in the sheet *Weibull-dist*, we will introduce discrete values of daily V covering the aforementioned range, in order to build the table of **Mean Cubic Speed** as a function of the daily V. The same is presented at the right of the Weibull table. The mean cubic speed is the speed—assumed as constant during 24 h—that would produce the same energy density resulting from the Weibull distribution along a day. There are two equivalent mathematical expressions for the mean cubic speed:

$$V_{c.avg} = \sqrt[3]{\frac{\sum_{v=3}^{n} v^3 \cdot t_v}{24}}$$

where:
  v = discrete speed of the wind from the histogram (3, 4, 5...n = max. speed considered in the histogram) (m/s)
  t_v = number of hours in a day that the wind blows at the speed v

and

$$V_{c.avg} = \sqrt[3]{\frac{2E}{24 \cdot 1.22}}$$

where:
  E = energy density (Wh/m²·day)
  24 = hours in a day
  1.22 = standard density of air usually assumed for calculating E (kg/m³)

Please note that, for practical purposes, the energy in the interval of speed 1–3 m/s is irrelevant, since no turbine produces energy in such range of winds. Even turbines having V_{cut-in} < 3 m/s, will produce no relevant amount of energy below 4 m/s.

Now we can generate the sheet *Simulation*, which is a copy of *clean avg 2009*, with two additional columns: the mean cubic speed corresponding to the day's average speed, and the daily energy density. The mean cubic speed results from the corresponding table. In our version of spreadsheet, it can be retrieved automatically using the built-in function SEARCH. VERTICAL, and then copying downwards from the first row to the last one. Thus, we have determined the probable mean cubic speed and the air density corresponding to each day. Calculating the daily useful energy density is straightforward.

10.4.2.3 Finding the Optimum Turbine

The sheet *Electricity Production* has two tables: the first one is the required energy for pumping in each month, resulting from the calculations in the former paragraphs. The second table is the total energy produced in each month, resulting from multiplying the monthly energy density—from the sheet *Simulation*—by the area and C_p of the desired turbine (from the manufacturer's catalog). Checking different turbines with different sizes, C_p and installation costs, we can now find out the most suitable solution. See the sheet *Cash-flow*, where the evaluation method adopted is that of the Net Present Value (NPV), a built-in function of the spreadsheet.

N.B.: The costs and tariffs presented in the sheet *Cash-flow* are just indicative; the choice of 5% interest rate is arbitrary. The scope of the said sheet is only didactic.

10.4.3 Tailoring a Wind Pumping System for a Given Context

Shadoof or *shaduf* is the Arabic name of the water pole or counterpoise lift, a very primitive water-lifting device (Figure 10.18), still in use in many rural areas of Northern Africa and Middle East.

Within an international cooperation project, our mission will be designing a simple and cheap windmill to replace muscular labor. The irrigation water must be lifted 1.3 m from a canal to irrigate nearly 1.5 ha of vegetable garden and orchard. Currently, the farmer must work around 4 h/day for lifting the necessary water for his crops with the existing *shadoof*. According to the Global Wind Atlas, the annual power density at 50 m is 75 W/m². The roughness height is $h_0 = 0.3$ m. The annual average wind speed at 50 m is $V = 4$ m/s.

10.4.3.1 Determining the Necessary Power for the Windmill

The farmer does not know exactly how much water his crops need. He gives just the indication that he works nearly 4 h/day lifting water with the *shadoof*. We know that the average power that a man can maintain during a 4-h shift is at least 75 W (see Chapter 1, Section 1.1). The efficiency of a *shadoof* is relatively high, because it has very little friction, but its operation requires some movements that are not connected directly to lifting the water, for instance, tilting the full bucket and pouring the water in the upper irrigation canal. We can conservatively estimate the overall efficiency in about 80%. Hence, the net energy employed for lifting water, and the corresponding lifted volume in a 4-h shift, q, are:

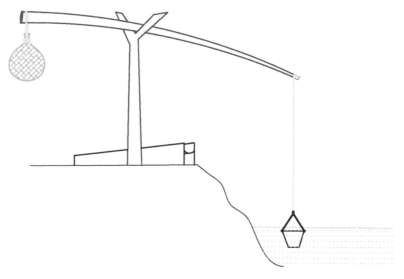

FIGURE 10.18
Sketch of a *shadoof*.

$$E_{\text{manpower}} = 0.8 \cdot 75 \text{ W} \cdot 4 \text{ h} = 240 \text{ Wh} = 864 \text{ kJ}$$

$$E_{\text{potential}} = \rho \cdot g \cdot h \cdot q = 1{,}000 \text{ kg/m}^3 \cdot 9.8 \text{ m/s}^2 \cdot 1.3 \text{ m} \cdot \text{qm}^3 = 12.74 \cdot q$$

$$E_{\text{potential}} = E_{\text{manpower}} \rightarrow q = \frac{864 \text{ kJ}}{12.74 \text{ kJ/m}^3} = 67.8 \text{ m}^3$$

We can cross-check the estimation above with the help of Table 10.4. According to the same, the amount of water to irrigate 15,000 m² of vegetable garden is in the range 45–75 m³/day. We are going to assume the maximum of such range as design parameter, in order to ensure enough watering in summer without the farmer having to integrate the wind pump with additional labor at the *shadoof.*

10.4.3.2 Choosing and Sizing the Wind Turbine

The chosen wind turbine is the *Savonius* type, because our customer, an international cooperation NGO, wants such solution. The reason for their choice is more ideological than technical: they want to employ locally available carpenters and blacksmiths, and consider that *Savonius* turbines are the easiest to build. The pump could be either a piston, or a diaphragm, or a bellow pump, eventually coupled by means of pulleys and belt, if a speed reduction should be necessary. The maximum foreseen height of the turbine's support structure is 6 m, because of construction limitations. It is possible to cast on site one or more concrete slabs to form the bases for both the structure and the pump.

The first step is to estimate the wind speed at 7 m height, which will be roughly the center of the turbine. We start with the data at 50 m height: with the help of the file *Savonius-pump.xlsx*, in the sheet *Weibull*, we are going to determine the coefficients of a Weibull distribution that yields 75 W/m² with an average speed of 4 m/s. A reasonable solution, considering that the place has weak winds but low rugosity, is $k = 1.8$ and $F = 0.965$. With such values, the wind speed that corresponds to the peak of the energy density is 7 m/s, and the probability that the wind speed exceeds 14 m/s is practically null. The opposite extreme is $k = 1.3$ and $F = 0.85$, but such values would "flatten" the probability distribution, shifting the speed range up to 19 m/s, a wind speed that appears too high for being probable in a subtropical place. Now we can calculate the corresponding wind speed at 7 m height for each of the speeds in the interval considered, using the logarithmic formula introduced in Chapter 8, Section 8.5.

The table at the right in the sheet *Weibull* contains then the resulting wind speeds at 7 m height, and their statistical time distribution is the same as the wind at 50 m height. From the said table we can observe that the energy density for $v < 2.4$ m/s is irrelevant, so we will assume $V_{\text{cut-in}} = 2.4$ m/s. The average power density for $v > 2.4$ m/s is then 16.51 W/m². The peak of the energy density function corresponds to 4.3 m/s while the mean cubic speed is 3 m/s. If we design our *Savonius* rotor to have $V_{\text{cut-in}} = 2.4$ m/s, it will be able to extract 95% of the wind power available in the site.

The C_P of a *Savonius* rotor is 0.3 and we can conservatively estimate the overall efficiency of the pump and transmission as 80%. The average wind power necessary to lift 75 m³ of water to 1.3 m head in 24 h, with the said efficiency is:

$$P_{\text{wind}} = \frac{1,000 \ \text{kg/m}^3 \cdot 9.8 \ \text{m/s}^2 \cdot 1.3 \ \text{m} \cdot 75 \ \text{m}^3}{0.3 \cdot 0.8 \cdot 86,400 \ \text{s/day}} = 46.1 \ \text{W}$$

Since the useful energy density of the site at 7 m height is 16.5 W/m², our *Savonius* rotor must then have 2.8 m² of exposed area, S. The turbine will operate at its maximum C_P when the speed is 3 m/s, and the power at its shaft will be 13.83 W.

It is easy to demonstrate that the amount of material necessary to build the semi-cylinders of a *Savonius* rotor of given S is independent of its H/D ratio, while the amount of material necessary to build its base, top, and eventual intermediate plate is proportional to the square of its diameter. Therefore, the proportions of the rotor can be arbitrary, but there is a slight convenience in maximizing its height, H. On the other hand, when *Savonius* rotors must drive a water pump it is convenient to maximize the diameter, since the torque is proportional to it. The usual construction of *Savonius* rotors employs dismissed steel or plastic drums, in which case the rotor's diameter is nearly twice the drum's diameter, and the required height results from stacking as many drums as necessary. In our case, no steel drums are available on site, so the semi-cylinders will be built using steel plates and we have absolute freedom to choose the rotor's proportions. Since we assumed that the center of the turbine will be at nearly 7 m, and the maximum height of the base can be 6 m, then the turbine must be at least 2 m high. Hence its diameter will be $D = 1.4$ m.

10.4.3.3 Designing the Pump and the Rotor to Match Each Other

In order to keep the construction as simple and cheap as possible, our goal is designing the rotor and the pump in such a way that there is no need of a reduction gear to couple each other. In other words, we are going to design the pump's bore and stroke in such a way that the rotor is able to drive it with maximum efficiency at $V = 3$ m/s, and furthermore to work with good efficiency at $V = 4.3$ m/s, because such wind speed produces the maximum amount of energy during the year. At the end of the process, we must check that the torque at $V = 2$ m/s is enough to ensure the start of the rotation.

We must recall from Chapter 6, Figures 6.3 and 6.6, that all *Savonius* rotors have the following characteristic coefficients: $\lambda_{\text{opti}} = 1$, $C_{P \text{ opti}} = 0.3$, $C_{M \text{ opti}} = 0.3$. From the definition of λ (Chapter 2, Section 2.4.3.2)

$$\lambda = 1 = \frac{\omega R}{V} = \frac{\pi DN}{60V} = \frac{\pi \cdot 0.7 \cdot N}{60 \cdot 3}$$

we can deduce that $\omega = 4.28$, $N = 81.85$ R.P.M, $n = 1.36$ turns/s.

Recalling the relationship between M and C_M,

$$M = \frac{\rho}{2} \cdot C_m \cdot R \cdot S \cdot V^2$$

the calculation of M is straightforward:

$$M = \frac{1.22 (\text{kg/m}^3)}{2} \cdot 0.3 \cdot 0.7 \text{ m} \cdot 2.8 (\text{m}^2) \cdot 3^2 (\text{m}^2/\text{s}^2) = 3.23 \text{ (Nm)}$$

We need to lift in average 75 m^3 per day, i.e.,

$$Q = \frac{75 \text{ m}^3}{86,400 \text{ s/day}} = 0.00087 \text{ m}^3/\text{s} = 0.87 \text{ l/s}$$

We are going to employ a double effect pump, both because of its higher efficiency and because it will absorb torque more regularly from the rotor, avoiding the need of a compensation mechanism and facilitating the start of the turbine at the lowest wind speed. Standard plastic tubes are available in the nearby town. It is relatively easy to build a pump with plastic tubes according to the sketch shown in Figure 10.19. We only need to calculate which the optimum section and stroke of the pistons are.

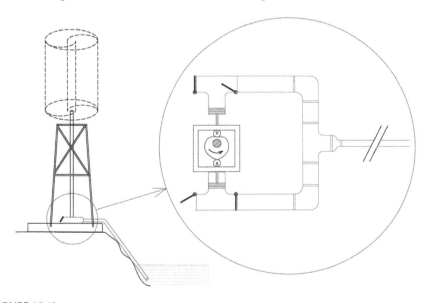

FIGURE 10.19
Two single effect piston pumps with opposite movement, linked by a yoke cam, behave as a double effect pump having twice their single displacements.

From Figure 10.19, we can deduce that each turn of the shaft corresponds to two strokes, so a double effect pump has $k = 2$ without the need of a gear. During half of the turn, one of the pistons will suck water from the canal (i.e., it will perform the required mechanical work) while the other will expel the water aspired during its previous stroke (mechanical work is negligible, because the discharge is at the same level of the suction head, directly on a trough). During the next half of the turn, the movement inverts and the pistons swap their functions. The overall result is a continuous suction, a continuous torque absorbed from the shaft and an alternating discharge from each of the pistons. It is relatively easy to build the valves and pistons on site with plastic or metal discs and rubber from an old tire inner tube. For details about homemade pumps with plastic tubes, see Longenecker, J., *PVC Water Pumps*.

The force exerted by the connecting rod at each upstroke corresponds to the weight of the water column in the suction tube:

$$F = 1,000 \text{ kg/m}^3 \cdot 9.8 \text{ m/s}^2 \cdot s_s \cdot 1.3 \text{ m}$$

where s_s is the section of the suction tube. The choice of the suction tube is more or less arbitrary: minimizing the section will maximize the viscous losses, but at the same time will minimize the necessary force on the piston's connecting rod. The rule of thumb for a compromise between minimum section and acceptable frictional losses is choosing the tube's diameter such that the speed of the water in it is less than 1.5 m/s. In our case, we need at least 38 mm (1 ½"). Hence:

$$F = 1,000 \text{ kg/m}^3 \cdot 9.8 \text{ m/s}^2 \cdot 0.00113 \cdot 1.3 \text{ m} = 14.45 \text{ N}$$

It is necessary to consider also the friction force of the pistons against their cylinders' walls and the viscous and dynamic drag forces caused by the water flowing across the check valves and along the tubes. Such forces are complex to evaluate, since they depend not only on the quality of the pump's construction, but also on the flow rate, furthermore, they vary along the piston's stroke. As a first approach, we can estimate the mean overall friction forces as 20% of the lift force. Hence:

$$F' = 17.4 \text{ N}$$

Now, we must check if it is possible to couple the pump directly to the turbine's shaft otherwise, what the necessary multiplication ratio k is. In this example, we are going to perform analytically the graphic procedure explained in Section 10.3.4.1.

We want the straight line representing the power absorbed by the pump to pass through the maximum of the turbine's power curve at 3 m/s. Hence:

$$P_{pump} = \frac{\rho \cdot g \cdot h \cdot q \cdot n_{pump}}{\eta}$$

All data in the equation above are known, except the piston's displacement, q. Please remember that n_{pump} is the number of strokes per second, which is twice $n_{turbine}$.

On the other side, we know that $P_{turbine} = 13.83$ W and $n_{turbine} = 1.36$ turns/s. Hence:

$$q = \frac{13.83 \cdot 0.8}{1000 \cdot 9.8 \cdot 1.3 \cdot 2 \cdot 1.36} = 0.00032 \text{ m}^3$$

The second condition to comply with is the balance of driving and resisting torque. Since the force on the piston's rod is nearly constant in module throughout the full turn of the shaft, then:

$$M_{turbine} \geq F' \cdot a \rightarrow a \leq \frac{3.23 \text{ (Nm)}}{17.4 \text{ (N)}} = 0.186 \text{ m}$$

Now we need to determine the cylinder's diameter, in order to get $L \approx 2a$. Table 10.5 shows the resulting value of L for the standard piping sections available on site.

The tube with 38 mm diameter would provide a reasonable stroke for building a crank or a Scotch yoke driving a pair of pistons. On the other extreme, the tubes with 100 mm diameter provide a very short stroke, which opens the possibility of building a diaphragm pump driven by a cam and follower mechanism. The choice of the cylinder's diameter and reciprocating mechanism is more or less arbitrary within the limits defined in Table 10.5. There is some convenience in adopting the biggest diameter, in order to minimize the viscous losses caused by the check valves. The diaphragm pump is easy to build with a rubber membrane cut from an old inner tube of a tire. Even a stroke as long as 80 mm would provide a value of a well below 0.186 m.

Now we can calculate v_{cut-in}. Since the static coefficient of torque is 0.35 (Figure 6.6) then:

TABLE 10.5

Standard Plastic Tubes, Their Sections, and Resulting L When Employing Them as Cylinders for the Piston Pump

Std. D (mm)	s (m²)	$L = q/s$	$a = 1/2$
25	0.00049	0.652	0.326
38	0.00113	0.282	0.141
50	0.00196	0.163	0.081
75	0.00442	0.072	0.036
100	0.00785	0.041	0.020

$$M = \frac{1.22\,(\mathrm{kg/m^3})}{2} \cdot 0.35 \cdot 0.7\ \mathrm{m} \cdot 2.8\,(\mathrm{m^2}) \cdot v_{\text{cut-in}}^2\,\mathrm{m^2/s^2} = 3.23\,(\mathrm{Nm})$$

Therefore, $V_{\text{cut-in}}$ = 2.76 m/s, which is slightly higher than 2.4 m/s, but acceptable.

When V = 4.3 m/s, since the torque of a volumetric pump is almost independent of its speed, then the torque coefficient of the turbine must decrease:

$$M = \frac{1.22\,(\mathrm{kg/m^3})}{2} \cdot C_M \cdot 0.7\,\mathrm{m} \cdot 2.8\ (\mathrm{m^2}) \cdot 4.3^2\,(\mathrm{m^2/s^2}) = 3.23\,(\mathrm{Nm})$$

C_M = 0.146, which corresponds to λ = 1.46 (see Figure 6.6), therefore C_P = 0.21.

The turbine's power with such C_P is then 28.52 W. The flow rate will be 1.8 l/s.

10.5 Conclusions

Water is key for agriculture, and pumping water in the most sustainable way is crucial for coping with desertification and carbon emissions. Wind turbines are by far more sustainable than any other renewable energy technology, especially in developing countries, because they are easy to build with local materials and skills. It is important to remark once more that direct wind pumping, even with low efficiency self-constructed turbines, is generally more efficient than generating electricity with a wind alternator and then pumping with an electrical centrifugal pump.

There is no "one fits all" solution for pumping water with wind power. Depending on the socio-economical context and available infrastructure, the designer can choose between at least four families of wind turbines and six types of volumetric pumps, plus the commercial combination of electric fast turbine with electric submerged pump, i.e., 25 potential ways of solving each case. The choice contains of course much of personal discretion and sometimes the adopted solution may be not the optimum, but results from local constraints.

Bibliography

Centro Las Gaviotas/United Nations Development Program, *Installation Manual of the Gaviotas Double Action Tropical Windmill* (English), Orinoquia, Colombia. Free to download from http://www.centrolasgaviotas.org/docs/Gaviotas%20 Windmill.pdf.

Chaurette J., Chapter 4: Pump selection, sizing and interpretation of performance curves, *Pump System Analysis and Centrifugal Pump Sizing*, http://www.pumpfundamentals.com/pump_book, free online book, http://www.pumpfundamentals.com/download/book/chapter4.pdf. Free online calculators in the educational site www.pumpfundamentals.com/.

Dartnall J., Milne-Home W., and Reid A., *An Efficient Piston Pump for Low-Yield Aquifers*, University of Technology Sydney, Australia, 2012.

Fraenkel P.L., *FAO Irrigation And Drainage Paper 43, Water lifting Devices*, Food and Agriculture Organization of the United Nations, Rome, 1986. Free to download from http://www.fao.org/docrep/010/ah810e/AH810E00.htm#Contents.

Le Gourière D., *L'Énergie Éolienne – Théorie, conception et calcul practique des installations*, deuxième édition, Eyrolles, Paris, 1982.

Longenecker J., *PVC Water Pumps*, ECHO Technical Note, North Fort Myers (FL), USA, 2010. Free to download from www.echonet.org.

Lüdecke H.J. and Kothe, B., *Water Hammer*, KSB AG Communications, Halle, Germany, 2015. Free to download from https://www.ksb.com/blob/7228/b03ed4dd6aa0139a876090d66fe3b9f2/dow-know-how1-water-hammer-data.pdf

Rosato M., *Progettazione di impianti minieolici*, Multimedia course, Acca Software, Montella, Italy, 2010.

Rosato M., *Diseño de máquinas eólicas de pequeña potencia*, Editorial Progensa, Seville, Spain, 1992.

11

Unconventional Wind-Driven Machines

11.1 Introduction

Nothing seems to stimulate the imagination of "inventors" around the globe like capturing wind's energy. The Internet is plagued of blogs, sites of start-ups, and homemade videos presenting "the ultimate technology" for wind power, but despite of the many claims most concepts remain as sketches or photorealistic renderings, few become working prototypes, fewer become commercial products, and none as far as the Author knows has ever been able to beat 3-bladed horizontal axis wind turbines (HAWTs) by price and performance. The reason is simple: no matter what "inventors" claim, Betz's limit is a law of Nature nobody can break, and fast HAWTs are the physically feasible machines that most closely resemble Betz's ideal actuator disk. Furthermore, 3-bladed HAWTs require the least amount of material for a given power and hence they offer the lowest cost per installed kW than any other wind energy converter.

In spite of evident flaws in their designs and much hype about their performances, some of the so-called innovative wind energy converters manage to get conspicuous funding from big companies or from public institutions like the European Commission. Such financial endorsement does not mean that the new concept actually has any concrete advantage on 3-bladed HAWTs, but at the same time is a painful show of wasted resources in a critical moment in which our planet requires effective solutions to avoid climatic catastrophes. Such wasted resources could have been employed better if decision makers were just keener in searching the Internet to check if "new" technologies are really "new," or if they had a better understanding of wind power technology and its long history of test and failure. The following sections will provide the reader an overview of the most recurrent "alternative technologies" proposed in the wind power market, together with an explanation of why the same have no future or are not even likely to work.

11.2 High Altitude Concepts

Since wind power is a function of the cube of the speed, and the speed and steadiness of wind increases exponentially with the altitude, it is evident that placing the turbine as high as possible will require a smaller turbine and hence lower cost for a given amount of energy produced along the year. There are some flaws in such simplistic reasoning:

- Placing a turbine at a high altitude does not necessarily cost less than building a bigger turbine on ground.
- The exponent of the wind speed variation with altitude depends on the area's roughness. A conventional 3-bladed rotor placed at 100 m height in an offshore location where the roughness class is 0, can have higher energy productivity than a flying rotor at 500 m height over land with high orographic roughness, like forests or mountains.
- Flying wind turbines create a legal problem: a no-flight zone must be declared where such devices are deployed.
- If the cable that holds the device breaks, or if the device itself loses its lift for any reason, it can fall on people. Hence, such systems should be placed in desert areas, but then the construction cost and the losses of the electric line that conveys the energy to the final users can be of the same order of magnitude, or higher, than the prospected increase of productivity or investment savings in comparison with conventional turbines.

High-altitude wind energy production technologies can be divided in two main categories: "flying" or "airborne" generation and "ground" generation. Technologies belonging to the first group aim to bring alternators to high altitude by means of an aircraft and the generated energy is brought down to the ground by means of a cable, which also works as a tether to maintain the aircraft linked to the ground. The main players adopting this layout are based in North America. In "ground" technologies, mechanical energy is somehow collected in altitude (usually by aircrafts or "kites") and transferred through one or more lines to the on-ground alternators. The main players adopting this layout are based in Europe (Corongiu, 2012).

The features and weak points of current high-altitude wind power concepts are discussed in the next sections.

11.2.1 Kites

The most conspicuous project is probably the Italian KiteGen (www.kitegen. com), funded in part by the European Union. According to the company, 12 million euros had been spent by December 2015, of which nearly 10% of public money. Kites can work in "yo-yo" configuration, like the KiteGen, or in circular path configuration. In the first case, the kite is set at an attack angle that maximizes the pull on the line. The pull force unwinds a reel, which is connected to an alternator that produces then electricity. When the line reaches its total length, the kite's attack angle is changed in order to minimize its drag, and the alternator is reversed to work as motor, rewinding the reel. The company claims that the energy necessary to rewind the reel is much smaller than the energy produced by the kite's pull, and that the cost per installed MW is much lower than that of conventional wind turbines. The weak point seems to be the reel: the cables employed to hold the strong pull of the kite are made of Kevlar, which is somehow abrasive for the reel and for the other cables, and the pull also loads heavily the bearings of the latter, so the maintenance cost of the on-ground installation is potentially high.

The "competitor" is the German SkySails concept (www.skysails.info). The same works in "yo-yo" configuration, although following a different path than the one proposed by SkyGen.

The reader already knows from the theory explained in the former chapters of this book that the efficiency of a wind power system depends directly on the lift/drag ratio of the conversion device. Kites working in "yo-yo" configuration resemble more a *Savonius* rotor than an "actuator disk," so they most probably feature low conversion efficiency. SkySails claims an "energy density" of 1,800 W/m², almost double of the one attributed to offshore wind farms, but does not support such claim with any evidence.

KiteGen also proposes the *Carousel* concept, in which several kites are connected to a big wheel and describe a nearly circular path. In this case, according to the company's claims, the system behaves like a turbine with "virtual diameter" of several kilometers. The implicit low aerodynamic efficiency of a kite compared to an airfoil is then compensated by an exposed area at high quota that no conventional turbine could ever reach. The concept is so complex and the dimensions of the proposed system are so pharaonic (60 GW power), that it is unlikely that it will be actually built.

11.2.2 Blimps

Aerostatic buoyancy is an intuitive and straightforward concept on how to replace a pole or support structure to hold an alternator at high altitude. Balloons can be produced in any size, and do not fall to ground when there is no wind, so this concept could lead potentially to the construction of simple small-sized turbines. There are two ways in which aerostatic buoyancy can replace the support structure of a wind turbine:

a. A streamlined balloon or airship carries a three-bladed HAWT. This concept has been published in books and magazines since at least 1970. As far as the Author knows, no single unit has ever been built with such configuration. Altaeros Energies (http://www.altaerosenergies.com) has built a more sophisticated prototype of a blimp, which is composed of several torus joined together to form a duct, housing a small wind turbine inside.

b. The balloon itself is a turbine. The company Magenn built a prototype of a blimp filled with helium, which was claimed being able to swirl at 180 m height while tethered to the ground. The shape of the blimp was that of a *Savonius*-like rotor wrapped around a central cylinder. The lift is a combination of aerostatic and Magnus effect forces. The founder of Magenn, Fred Ferguson, built the first Magnus effect airship in 1980 and announced in 2010 that the company was about to start selling their flying rotor. As for November 2017, the official site (www.magenn.com) seems to be offline and the last references to the company date back to 2012.

The weak point of aerostatic-borne wind turbines is their size. They may theoretically provide higher power for a given diameter than conventional wind turbines, but they need huge volumes filled with helium to be able to float in the air. Big volume means high area faced to the wind, hence a horizontal force that creates a moment to the anchor point on ground, which in turn will lower the height at which the balloon floats. The cost of the anchoring, tether cable that also transmits power and blimp are most probably higher than the cost of a conventional turbine of the same rated power, with bigger rotor, placed on ground.

11.2.3 Autogiros or Flying Electric Generators

An autogiro is a kind of kite in which a free-turning wind turbine produces the lift force. The turbine's area faces wind with a certain angle, so the effective exposed area is small compared to the rotor's area. This means that in a Flying Electric Generators (FEG), part of the wind's energy is effectively converted into electricity, and transmitted to ground along the tether cable, and the remaining fraction is employed in keeping the FEG at the desired altitude. The interesting aspect of FEGs is that the angle between the rotor's plane and the wind direction can be controlled easily, allowing to keep the machine in operation even in very high winds. The Australian researcher Bryan Roberts built a prototype with two 4.5 m diameter counter-rotating rotors. The craft takes off like a helicopter, with both rotors powered via the tether cable until it reaches the desired height, then the angle of attack is increased and the current is switched off. At this point, the wind maintains the rotors turning and generates enough lift to keep the craft in position. The company Sky Wind Power www.skywindpower.com claims that FEGs can produce twice the energy of a ground-based turbine of the same power, because winds at high altitude are steadier and blow a larger number of hours per year. In other words, the capacity factor of FEGs is nearly twice the capacity factor of ground-based turbines of the same rated power. The capacity factor is the ratio between the energy effectively produced by the turbine in one year and the energy it would produce in the same time if running constantly at rated power. The company has calculated the capacity factor for some localities in the United States, at 4,500 and 10,000 m height. It seems unlikely that an FEG can ever reach such altitudes, because the tether and transmission cables would need to be at least twice long to account for the sag and downwind displacement, with consequent problems in holding their weight and ensuring flight stability.

11.2.4 Tethered Aircraft

The most famous concept is the Makani tethered airplane (https://x. company/makani/). The first prototype was a sailplane model with 5.5 m wingspan carrying two turbines of 5 kW each. The current version is very similar in shape to a sportive sailplane, but with eight wind turbines mounted on it. The wingspan is 26 m and the rated power is 600 kW. The tether cable is 500 m long. The complexity of the control and operation described in the company's website seems to be higher than those of any conventional ground-based wind turbine.

11.3 Claims of Efficiency Higher than Betz's Theorem

Although Betz's theorem is a law of Nature, similar to Carnot's theorem of thermal machines, the number of "inventors" and some manufacturers claiming efficiencies higher than the theoretical limit, or suspiciously close to it, is very high. Internet and the general lack of scientific preparation of some mass media journalists contribute to diffuse such hypes. There are two types of "higher-than-Betz" concepts: those in which intriguing machines would produce energy based on working principles that escape the concept of "turbine" (see Section 11.5), and those in which the manufacturer simply considers a smaller exposed area than the actual, which gives the impression of a higher efficiency. The norm IEC 61400-12-1:2017 specifies a procedure for measuring the power performance characteristics of a single wind turbine and is applicable to the testing of wind turbines of all types and sizes connected to the electrical power network. In addition, this standard describes a procedure to determine the power performance characteristics of small wind turbines (as defined in IEC 61400-2) when connected to either the electric power network or to a battery bank. The procedure can be used for performance evaluation of specific wind turbines at specific locations, but equally the methodology can be used to make generic comparisons between different wind turbine models or different wind turbine settings when site-specific conditions and data filtering influences are taken into account. In general, companies claiming higher-than-Betz efficiencies of their products never present any evidence of independent tests performed in compliance with the said norm.

The following examples illustrate two cases: the first one of a company claiming that their product has higher efficiency than a horizontal axis wind turbine because "it has non tip speed" and the second one in which the manufacturer computes the efficiency on the base of the rotor's diameter, instead of considering the whole system of rotor plus Venturi diffuser.

11.3.1 Saphonian 3D-Oscillating Membrane

The Tunisian company Saphon (www.saphonenergy.com) promotes a bladeless wind converter consisting of a concave membrane that, according to their claims, captures all the energy of wind like a ship's sail—which by the way, is a false statement: a ship sail needs a difference of speed between both sides, otherwise it could not generate aerodynamic pressure. The company claims that such concept converts the kinetic energy of the wind into electricity at an efficiency twice that of Betz's theorem (*sic!*), i.e., 112%, which implicitly means that the machine violates the law of energy conservation. The website presents as "proof" just a video of a test performed with a set of ventilators at the roof of a building, and an object that looks like an umbrella moving in the air stream. They do not present any numerical result, not even a curve of power versus wind speed, nor evidence of third parties having tested their product independently.

11.3.2 Ducted Turbines

Most ducted turbines manufacturers claim efficiencies higher than Betz's theorem based on considering the area of the rotor only, which is in general a fraction of the total area exposed to the wind. A concentrator and a diffuser are equivalent from the point of view of fluid dynamics: the first is a funnel that gathers wind in a big area and conveys it to a smaller area, increasing then the speed across the latter. The second makes the opposite: it causes the air downstream of the turbine to expand, lowering its density and pressure, which in turn increases the speed across the rotor. Although being static parts, both the diffuser and the concentrator interact with the wind, so the area that needs to be accounted for the calculation of the efficiency is the total area of the rotor plus the maximum area of the ducting, and not only the rotor's.

11.4 Old Technologies Pretending to Be New

11.4.1 Spiral Surface Rotor

The Dutch company The Archimedes (http://dearchimedes.com/) and the American company Windinstrument (http://windstrument.com/) promote small turbines with solid rotors featuring spiral surfaces. Both claim that their design is innovative and aerodynamically superior to conventional turbines. The truth is that both designs are nothing but variations of the Dumont turbine, built and patented by the French engineer Alexandre Dumont at the end of the nineteenth century. The patent has the number 400065A, and the reader can check it in http://www.google.com/patents/US400065. A description of the Dumont turbine was published in the NASA Technical Translation TTF-16201 by R. Champly (1975), page 125, figure 84, which quotes a French publication of 1883. Apart from the inexact claim about the innovativeness of their products, the companies seem to ignore the basic relationship between specific speed, solidity, and coefficient of power. Furthermore, if we apply the theory of the blade element to a helicoid's surface, the differential section turns to be a flat plate instead of an aerodynamic profile. A flat plate, even at very small angle of attack, has lower C_z/C_x ratio than an airfoil, hence, such rotor has lower C_P compared to a conventional rotor with profiled blades.

11.4.2 *Savonius*-Like Rotors

The concept of the *Savonius* rotor is so intuitive that it seems to inspire architects and designers throughout the world to create their own variations, most often as integrations in buildings or urban spaces. The following are just two examples of the vast offer of *Savonius*-like products present in the market.

11.4.2.1 The Wind Tree

The French company Newwind (www.newwind.fr) proposes since 2011 a structure that resembles a tree with its branches, where the "leaves" of the tree are small *Savonius* rotors in molded plastic. They also propose the integration of batteries of such molded *Savonius* rotors into buildings, attached to the walls and roof. The company has won several awards to its innovative design. The product is undoubtedly original and aesthetically pleasant, and the company's marketing efforts focus on satisfying a "green" niche of consumers, for instance, by announcing that the plastic rotors are reinforced with hemp fibers.

11.4.2.2 Twisted *Savonius Rotors*

Figure 11.1 shows just one example of an Italian model, among dozens of similar variations that can be found in Internet. No matter if the rotor

FIGURE 11.1
A model of twisted *Savonius* rotor proposed by the Italian company *it-energy*, the day of its commercial launch. Photo by the Author.

features straight or twisted semi-cylinders, its aerodynamic efficiency does not increase with the shape or number of the same. The manufacturer of the model depicted claimed efficiency very close to Betz's limit. As far as the Author knows, the model is no longer in production.

11.4.3 Variations of the Pannemone

The panemone is one of the oldest and least efficient types of windmill. Some authors claim it was invented in China, others say it was invented in Persia. Nashtifan, at the East of Khorasan province of Iran, is an ancient city where some ancient windmills still work, although they are not exactly panemones, but ducted paddlewheels. In any case, the original panemone's rotor was most probably built with sails supported by wooden frames. Figure 11.2 shows the general operation principle of the flapping vane panemone. Its main disadvantages are the noise of the vanes clapping and a jerky rotation, caused by the variable momentum of inertia as the gravity center of each vane displaces along the circular path of their hinge points.

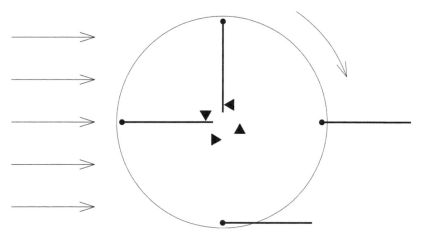

FIGURE 11.2
Schematic operation principle of the flapping vane pannemone.

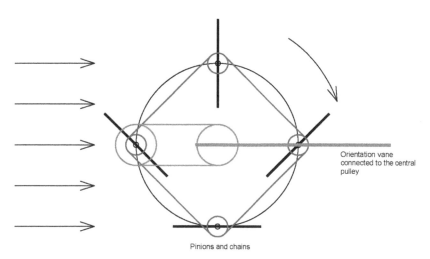

FIGURE 11.3
Schematic operation principle of the Cycloturbine.

11.4.3.1 The Cycloturbine

The Cycloturbine is a vertical axis wind turbine similar to the panemone, in which the vanes change their exposed area to the wind by means of some kind of epicycloid mechanism, oriented to the wind by a vane. The oldest patent known to the Author dates from 1863, and was meant as waterwheel (US Patent 38383 A, https://www.google.ch/patents/US38383). Figure 11.3 shows its operation principle.

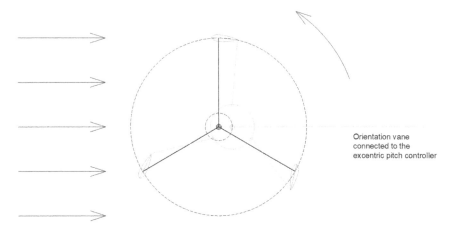

Orientation vane
connected to the
excentric pitch controller

FIGURE 11.4
Schematic operation principle of the Giromill.

11.4.3.2 The Giromill (a.k.a. Gyromill, a.k.a. Cyclogiro)

The difference between the Cycloturbine and the Giromill is that the first relies on differential drag forces to create torque, while the second can be defined as a Darrieus turbine with varying pitch along the blade's path (Figure 11.4).

11.4.3.3 The Vertical Axis Disc Turbine

This one is an ingenious homemade variation of the panemones, among the many featured in YouTube by a myriad of "inventors." The video is published in https://www.youtube.com/watch?v=sCSnq8tksgs. We are going to analyze its performances in Section 11.6.

11.4.3.4 The Costes Wind Motor

This simple pinwheel is described in the already cited NASA Technical Translation TTF-16201 by R. Champly (1975). It consists of six V-shaped troughs fixed to central shaft, as shown in Figure 11.5. The drag coefficients of a V-shaped profile exposed to the wind on its concave and convex sides are 2.22 and 1.45, respectively (see Figure 12.1). The calculation of the C_P is straightforward and is left to the reader.

11.4.3.5 The Lafond Turbine

This turbine and its performances are described in detail in the already cited NASA Technical Translation. It can be defined as a hybrid panemone, in the sense that its driving forces are caused partly by differential drag, partly by

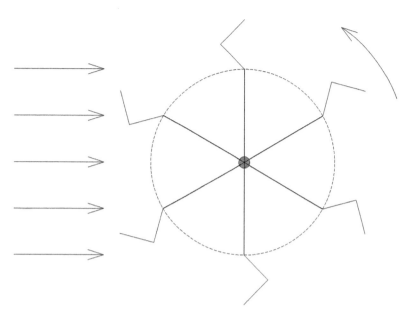

FIGURE 11.5
The Costes wind motor.

aerodynamic lift. Figure 11.6 shows the schematic disposition of its blades and airflow. Figure 11.7 shows the power output curves derived from the data published in the already cited NASA document, compared to the data declared by the manufacturer Zephyr.

11.4.4 Einfield-Andreau Pneumatic Gear

The Einfield-Andreau concept consists of a 2-blade HAWT with hollow blades and hub. The hub is connected to a tubular support structure holding a small diameter turbine inside. The idea is that the wind causes the rotor to turn, and the centrifugal force expels the air though the blades' tips, in a similar way to a centrifugal fan. The hub is then at a lower pressure than the outside, aspiring air inside the tubular support structure. Such air moves at high speed because the section of the tower is much smaller than the area exposed to the wind, so the scope of such configuration is to avoid placing the generator and gearbox at the top of the support tower. It is easy to demonstrate that the overall energy conversion efficiency is low, because of too many transformations: linear kinetic energy of the wind into rotating kinetic energy of the rotor; kinetic energy into pressure difference; pressure difference into linear kinetic energy of the air aspired into the hollow support structure—with the unavoidable friction losses; then linear kinetic energy again into rotating kinetic energy of the small inner turbine—assumed coupled directly to the generator—and finally rotating kinetic energy into electricity.

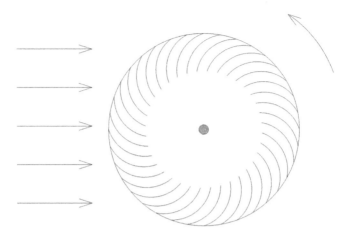

FIGURE 11.6
Schematic operation principle of the Lafond turbine.

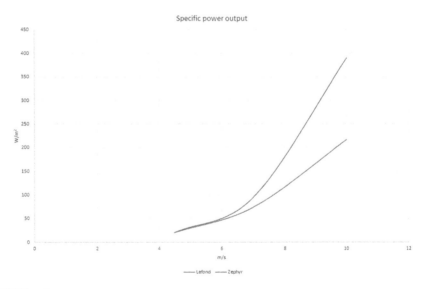

FIGURE 11.7
Performance of the original Lafond turbine compared to its modern equivalent manufactured by the company Zephyr. Numeric data from the cited NASA technical translation and the manufacturer's catalog, respectively. Data elaboration and graphics by the Author.

11.4.5 Darrieus Turbine with Its Axis in Horizontal Position

There is not much apparent sense in placing a vertical axis wind turbine (VAWT) with its axis in horizontal position. Some architects and designers have proposed such concept as a way of capturing energy from the vortexes caused by the edges of buildings. The main drawback of the idea is that the turbine is installed in a fixed position and is able to produce energy only when the wind blows in the suitable direction. This means that a building having rectangular or square base would require four of such turbines, one for each of the building's top edges. The Italian company Enatek (www. enatek.it) proposes a kind of Darrieus H rotor placed horizontally. Although the company claims that the *Venturbine®* provides more power than a similar turbine of the same size, because it exploits the "wind concentrator effect" of vertical walls, most probably such concept tends to overcome bureaucratic barriers rather than to provide an efficient solution. Italian norms specify that "turbines having 1 m diameter and outstanding less than 1.5 m from the building's envelope" are exempted from a series of bureaucratic procedures of evaluation of the visual impact. A conventional wind turbine with 1 m diameter has less than 1 m^2 of exposed area, and with the typical wind speed urban areas would never be able to produce more than 100 W output. A turbine like the Enatek, placed along the edge of the roof of a building being, for instance, 10 m long, would present 10 m^2 to the wind—but only when the wind comes in the right direction. Hence, it could yield 1 kW power, but still benefiting from the bureaucratic simplifications.

From an aerodynamicist's perspective, one could argue that the airflow near the edges of building's roofs is too turbulent and features a speed gradient too high for any turbine to extract energy in an efficient manner. Consequently, the C_p will be smaller than that of the same turbine placed with its axis in vertical position in an undisturbed wind. From the potential user's perspective, the turbine appears too bulky—and presumably too expensive—for a very little power output.

11.5 Non-Turbines

Inventors' ingenuity has produced a large number of "wind converters" that cannot be considered turbines. The facility to create photorealistic models and diffuse them in Internet has led many common people, some incautious investor, and unfortunately public bodies too, to believe that such concepts are disruptive innovations with high market potential. Until now, none has proven to be more energy efficient and economic than conventional 3-blade HAWTs, though some could have competitive advantages in some special contexts.

11.5.1 Beating Wing

According to UNESCO's Handbook of Wind Pumps, this concept was developed by the German engineer Peter Bade and a prototype built in Berlin in 1975. The concept was meant to provide an easy-to-build and cheap

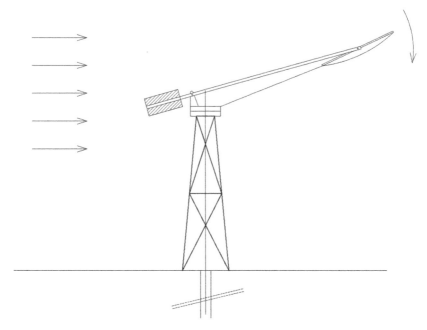

FIGURE 11.8
The flapping wing pump.

solution to pumping water from deep wells in arid and semiarid areas. Figure 11.2 is self-explaining. From an aerodynamic point of view, a flapping wing has some features that make it competitive, at least from a theoretical point of view, with *Savonius* and even slow HAWT. The power can be controlled by limiting the amplitude of the oscillation in high winds, thus reducing the swept area. Unlike turbines, the flapping wing does not induce three-dimensional flows, but only bidimensional, a feature that approaches more the ideal conditions of Betz's theorem than a rotor (Figure 11.8).

The Tunisian company **Tyerwind** proposes the "hummingbird" biomimicry concept. This wind converter features two wings made of carbon fiber reinforced plastic http://www.tyerwind.com/technology/#workingMachine. As per November 2017, the company declares that the performance curves will be disclosed once the tests are complete.

11.5.2 Linear Motion Rolling Blades

In this concept, a set of blades is fixed to two chains at each end, somehow resembling a Venetian blind. The blades follow a linear path perpendicular to the wind's direction, crossing the airflow twice. From a theoretical point of view, the idea is comparable to a HAWT with double rotor, with the advantage that the lift is uniform along the blade. From a practical point of view, building such a wind converter requires much more material than a HAWT, and the number of moving parts makes it less reliable and requiring more maintenance, as well as causing high frictional losses. Furthermore, such a system requires swinging the whole set of blades, chains, and rolls in order to face the wind.

11.5.3 von Karman Vortex Resonators

The general idea around this concept is that the wind will induce oscillations on a long cylinder or filament, because of von Karman vortexes. The energy conversion device can be either a piezoelectric generator or a coil and magnet linear generator. Supporters of such concept claimed that it is noiseless, which may be true in a very limited range of low wind speed, when the von Karman frequency is subsonic, but in such case the power produced is almost null. The biggest flaw of the concept is that the area swept by a cylinder oscillating with small amplitude is very small, so the cylinder should be very tall in order to produce a relevant amount of energy, which is in contradiction with the claims of low visual impact and low amount of construction material per unit of power. We already know that the wind speed is an exponential function of the height above ground. Hence, a very tall cylinder would be subject to a big wind speed gradient along its height, which means that the von Karman frequency will be different in different sections, leading eventually to the impossibility for the cylinder to enter in resonance and oscillate. Another contradiction in some of the claims that circulate in the Internet is that such devices would allow harvesting big power with low visual impact. Since the power is a function of the swept area, and the area swept by a vibrating rod is very small, then in order to produce relevant power a big number of units must be spread over a large land area, which is contradictory with the claim of low visual impact. The amount of material per unit of installed power will be also bigger compared to conventional 3-blade HAWTs, although the supporters of the concept claim the opposite. The Spanish company Vortex Bladeless is probably the most conspicuous example of the many ones that are promoting the vortex converter concept. The company claims huge performances and low cost but has produced no commercial unit so far (November 2017). In spite of the many logic flaws, contradictions and unsupported claims, Vortex Bladeless was funded with 1.32 million euros of public money from the European Union (http://cordis. europa.eu/project/rcn/204580_en.html). The project will end officially in May 2018, until then it is not possible to foresee what the results will be. The exercise proposed in Section 10.6.1 shows an optimistic theoretical estimation of the potential C_P and resulting size of such wind converter.

11.5.4 Delta Wing Vortex

Delta wings are flat surfaces capable of producing high lift when placed at angle in a fluid stream. The edges produce two strong counter-rotating vortexes. The low pressure at the center of each vortex is responsible of the high lift coefficient of the delta wing. Leftheriotis, G. and Carpenter investigated the idea of placing two counter-rotating turbines in the wake vortex of a delta wing in 1991. They concluded that such concept could have practical sense only for systems with less than 100 kW rated power and published the algorithm for designing a turbine rotor suitable for operation in non-uniform rotating flow. As far as the Author knows, no prototype has ever been tested.

11.5.5 Artificial Tornado

This concept has been presented in books on wind power since 1970. There are currently two philosophies on how to create an artificial tornado:

11.5.5.1 Wind-Induced Tornado

A cylindrical structure equipped with vanes deflects wind, creating a vortex. The vortex's low pressure attains air from the bottom and is conveyed through a Venturi-like duct. A small wind turbine placed in the point of maximum air speed extracts the energy. Such concept has a long list of disadvantages, first of all the huge amount of material per unit of installed power. As far as it is known, it has never been prototyped.

11.5.5.2 Heat-Induced Tornado

The creator of this concept is Louis Michaud, a retired engineer (https://www.popsci.com/science/article/2013-06/man-wants-power-world-tornadoes). The basic idea is to employ waste heat, for instance, from conventional thermal plants, to create a small tornado inside a high chimney, and recover its kinetic energy from a turbine placed at the top. The idea received some funding for prototyping in 2011, but as for November 2017, the Author has not found any evidence that it really works or can be economically competitive. The logic flaws in the idea of inducing a small tornado in a confined space are the following:

  a. the efficiency of conversion of thermal energy into mechanical energy is limited by the Carnot Theorem, hence, the efficiency in recovering waste heat at low temperature will be low;
  b. if relevant power is to be produced, the size of the heat exchangers must be big and consequently the size of the tube where the artificial tornado will form bill be proportionally bigger, so the installation cost will per unit of power will be too high;
  c. transmitting the rotary movement of air in the tornado chamber to an alternator requires a turbine similar to a pinwheel, which is intrinsically less efficient than a bladed rotor. It is probably more efficient to recover waste heat by induced draft (see solar chimney and related concepts in Section 11.5.7.1).

11.5.6 Magnus Effect: Flettner and Thom Rotors

Magnus effect is the lift force created by a fluid current on a rotating cylinder or sphere. A Flettner rotor (a.k.a. turbosail) is a cylinder rotating around its longitudinal axis (usually driven by a small motor) and a Thom rotor is a variant of the first, having discs uniformly spaced along its length. Magnus effect devices can reach extraordinarily high lift coefficients, which are proportional to their rotation speed. They feature high drag coefficients too (T. J. Craft, H. Iacovides, and B. E. Launder). We know from the theory on HAWT that the rotor's C_p is a function of the C_z/C_x quotient. Hence, it is clear that Magnus effect wind turbines present no aerodynamic advantage on conventional ones. According to Sedhagat, the C_p of Magnus effect turbines can be as low as 0.1. The Japanese company Mecaro produced a 10 kW model that features a spiral fin along each cylinder, but apart from a few videos posted in YouTube, there is no technical information available about it, not even in the corporate site.

11.5.7 Artificial Wind

The basic idea is to create wind by harnessing temperature differences from different sources, either solar power or residual heat from industries, mimicking Nature in small scale. The site www.solar-tower.org.uk proposes a collection of many concepts based on the said principle. They can be broadly classified in updraft and downdraft artificial wind systems. It is easy to demonstrate that such systems are not efficient: they are nothing but thermal machines running with very small temperature gradients. Air features low density and low heat exchange coefficient, hence, producing artificial wind requires huge exchange areas and extremely high chimneys for getting low power.

11.5.7.1 Solar-Induced Updraft a.k.a. Solar Chimney

A big area covered with greenhouses heats air, which is conveyed to a central chimney that induces updraft. A wind turbine placed at the base of the chimney extracts the energy from such air current. A prototype was built in Manzanares, Spain, and tested from 1982 to 1989. The chimney was 195 m high, 10 m in diameter, fed with hot air with 46,000 m² of greenhouse, and produced a maximum of 50 kW power.

11.5.7.2 Evaporative-Induced Downdraft

This concept is the mirror of the former: hot air in tropical regions would be cooled at the top of an evaporative tower, and the downdraft air motion would drive a turbine placed in the same tower. It has the same drawbacks of the solar chimney with an additional disadvantage: it is necessary to consume energy pumping water to the top of the tower.

11.6 Practical Exercises

11.6.1 Do Vortex Converters Have Any Potential Advantage on Wind Turbines?

From the video presented as "proof of concept" by a company supporting such concept, we can deduce that the amplitude of the oscillation of the converter's cylinder, A, is nearly $10°$ (0.174 rad). Calculate the height of the cylinder necessary to produce 4 kW with 10 m/s wind. Assume that the cylinder's diameter is 20 cm, the cylinder is perfectly rigid (no energy dissipated in deforming it) and that the wind speed gradient with height is negligible.

Solution

The company provides no information about the C_P of its product, so we cannot calculate the necessary exposed area directly.

Von Karman's equation allows to calculate the frequency of the vortexes detaching from a circular geometry:

$$f = \frac{0.198 \cdot V}{D}\left(1 - \frac{19.7}{\text{Re}}\right)$$

where:
  f = detachment frequency of the airflow (Hz)
  V = wind speed (m/s)
  Re = Reynolds number = $V \cdot D / \nu$
  D = diameter of the cylinder (m)
  ν = kinematic viscosity of air = 0.0000149 (m²/s)
  Hence:
  Re = $134{,}228$ and $f = 9.9$ Hz

At the moment of detachment of the airflow from the cylinder, the wind speed on one side is practically null (optimistic simplification), while on the opposite side the air is moving at 10 m/s, hence, the difference of static pressure between both surfaces of the cylinder can be deduced from Bernoulli's equation:

$$P_{\text{atm}} = p + \frac{\rho \cdot V^2}{2} \rightarrow \Delta p = 0.5 \cdot 1.22 \cdot 100 = 61 \text{ Pa}$$

Since we assumed no wind speed gradient, the mechanical work performed by a differential section of the cylinder at each cycle is:

$$dW = dF \cdot r \cdot 0.174 \text{ rad}$$

The force dF is the product of Δp by the differential area $D \cdot dr$. The work performed in a unit of time is f times the differential work. The differential power dissipated by the wind in a unit of time is then:

$$dP = f \cdot \Delta p \cdot D \cdot dr \cdot r \cdot A$$

The power dissipated along an oscillating cylinder of length L is then:

$$P = f \cdot \Delta p \cdot D \cdot A \int_0^L r\, dr \rightarrow P = f \cdot \Delta p \cdot D \cdot A \cdot \frac{L^2}{2}$$

Replacing the known values, we obtain:

$$4,000 = 9.9 \cdot 61 \cdot 0.2 \cdot 0.174 \cdot 0.5 \cdot L^2$$
$$L = 19.5\,\mathrm{m}$$

The area swept by the oscillating motion is then:

$$S = \frac{1}{2} A \cdot L^2 = \frac{0.174}{2} \cdot 19.5^2 = 33.1\,\mathrm{m}^2$$

Hence the C_p of the device is:

$$C_P = \frac{4,000}{\frac{1.22 \cdot 33.1 \cdot 10^3}{2}} = 0.198$$

Assuming that the C_p and dimensions calculated before are reasonably realistic: Does the vortex converter offer any concrete advantage on a conventional turbine?

According to the defenders of such technology:

- "The visual impact of vortex converters is smaller than that of HAWT."

  This is more or less true. A HAWT of the same rated power at the same conditions would require a rotor of nearly 4.6 m diameter plus the orientation vane, all placed on top of a structure at least 10 m high. One could argue that the main parameter of visual impact of any structure is its height. If the wind speed gradient is considered, then the cylinder assumed for our simplified evaluation of the C_p is no longer such. In order to produce the same von Karman frequency along its full length, the converter must be conic, with the biggest diameter at its top, where the wind speed is maximum. This increases the visual impact.

- "Building a vortex converter requires less material than a wind turbine of the same rated power."

Such statement is questionable. It is true that in a vortex converter the weight is at the base, so it would require less concrete foundations, but it is also true that, in order to recover as much energy as possible from the oscillating motion, the cylinder must be very rigid, otherwise part of the vortex's energy will be dissipated in deforming rather than displacing it. In order to be reasonably rigid, a very slender cylinder like the one assumed in our example must have a big material section or, conversely, its diameter must be increased, what means lower oscillation frequency, which in turn translates into lower power for a given length.

- "Vortex converters can produce energy in a wider range of wind speeds."

  Such statement seems much of hype. For a given geometry of the oscillating cylinder, the power depends on the frequency and on the difference of pressure, which are proportional to V and V^2, respectively; hence, the power output of the vortex converter is proportional to V^3, like for any other wind-driven machine.

- "Vortex converters are more silent than wind turbines."

  This statement is true if the diameter of the cylinder is big enough to ensure oscillations at frequencies lower than 20 Hz, which are inaudible for people. It is also true that subsonic frequencies can be perceived by the body's sense of tact, and cause some sort of inconvenience to people and animals. In particular, the frequency $f = 7$ Hz can be harmful for people, because it can bring the internal organs of the body into resonance.

11.6.2 Check the Maximum C_P of the Pannemone Presented in Section 11.4.3.3

Since all drag-driven wind converters have their maximum C_P when $\lambda \approx 1$, we are going to check the theoretical forces on one elements of the disc panemone. Figure 11.9 shows schematically the geometry, relative speeds, and resulting forces.

Solution

Assume that the turbine has three elements, like the one shown in the cited video. If we analyze just one of them along its full circular path, we will notice that during the "active" path, when it is moving downwind (nearly one-third of the circle, i.e., assuming that the element opens and shuts instantly) its exposed area varies sinusoidally from the angular position 60°–120°, being its drag coefficient constant and equal to 2.22 (see Figure 12.1). We will assume optimistically that the drag force along the remaining 2/3 of the circular path is that of a thin flat plate whose area is roughly 3S. The drag

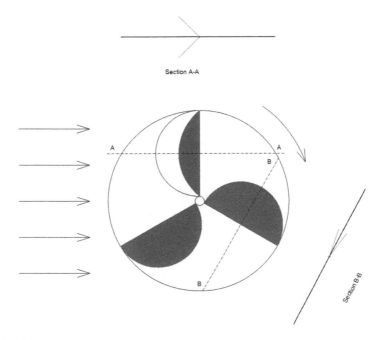

FIGURE 11.9
Sketch of the disc pannemone.

coefficient of a flat plate when Re > 10⁶ is a value in the range 0.004–0.014. Let's assume the most favorable case, 0.004.

Since there is only one exposed vane at a time, with the other two being folded for minimum drag, the power generated by the rotor along each turn is then:

$$P = M \cdot \omega = F \cdot \frac{r}{2} \cdot \omega = \frac{\rho}{2} \cdot v^2 \left[2.22 \cdot S \int_{\pi/6}^{2\pi/3} \operatorname{sen}\theta \, d\theta - 3 \cdot S \cdot 0.004 \right] \cdot \frac{r}{2} \cdot \omega$$

$$P = \frac{\rho}{2} \cdot v^2 [2.22 \cdot S - 0.012 \cdot S] \cdot \frac{r}{2} \cdot \omega$$

$$P = \frac{\rho}{4} \cdot 2.208 \cdot S \cdot v^3$$

The area swept by the rotor is nearly 4S. Hence, the approximate power coefficient is:

$$C_P = \frac{\frac{\rho}{4} \cdot 2.208 \cdot S \cdot v^3}{\frac{\rho}{2} \cdot 4 \cdot S \cdot v^3} = 0.276$$

Such C_P is in the same order of magnitude of a *Savonius* rotor's. Unlike the *Savonius* rotor, the panemone under analysis has a mechanism for opening and closing of the vanes. Such mechanism consumes part of the kinetic energy of the rotor, which was not accounted for in our simplified calculation. Hence, the real C_P must be lower than the calculated value.

Bibliography

Champly R., *Wind Motors: Theory, Construction, Assembly and Use in Drawing Water and Generating Electricity*, NASA Technical Translation TTF-16201, Washington, D.C. April 1975. Translation of a French book published in 1933.

Corongiu M., *High Altitude Wind On-Ground Energy Generation*, Forward Looking Workshop on Materials for Emerging Energy Technologies, EU Directorate for Research and Innovation, Brussels. 2012. http://ec.europa.eu/research/industrial_technologies/pdf/emerging-materials-report_en.pdf.

Craft T.J., Iacovides H., and Launder B.E., *Dynamic Performance of Flettner Rotors with and without Thom Discs*, University of Manchester, internal publication. 2011 http://www.homepages.ed.ac.uk/shs/Climatechange/Flettner%20ship/TSFP7-Flettner-Rotor-Paper4.pdf.

Leftheriotis G. and Carpenter C.J., Investigation of the delta wing vortex turbine system's suitability for large scale applications, *Wind Engineering* 15 (1), 16–24, 1991.

Leftheriotis G. and Carpenter C.J., Development of a turbine to operate in the vortex field generated by a slender delta wing, *Journal of Wind Engineering and Industrial Aerodynamics* 39 (1–3), 417–425, 1992.

Moran W.A., *Giromill Wind Tunnel Test Analysis*, Mc Donnel Aircraft report. October 1977. http://wind.nrel.gov/public/library/gwt.pdf.

Ragheb M., *Wind Power Systems*, University of Illinois online publication. 2017. http://www.ragheb.co/NPRE%20475%20Wind%20Power%20Systems/index.htm.

Rastogi T., *Wind Pump Handbook*, UNESCO Edition PGI/82/WS/17, Paris, 1982.

Rosato M.A., *Impianti Microeolici*, multimedia course, ACCA Software, Montella, Italy, 2012.

Sedhagat A., *Progress in Magnus Type Wind Turbine Theory*, conference paper, Energy Vol. 8. 2014. http://www.sedaghat.iut.ac.ir/energy-vol-8-wind-energy-808-progress-magnus-type-wind-turbine-theories.

12

Aerodynamic Characteristics of Blunt Bodies and Airfoils

12.1 Generalities

The purpose of this section is to provide a selection of formulas, tables, and elements of calculation that the reader will find useful for designing any type of wind turbine, including its support poles, mainstays, and other accessories. The data originate from different sources, reported in the bibliography.

12.2 Aerodynamic Characteristics of Extruded Profiles

The values of C_x shown below correspond to profiles with "infinite aspect ratio." In practice, they are applicable to extruded bars whose ratio between length and cross-dimension (diameter or length perpendicular to the airflow) is bigger than 10:1. The aerodynamic drag is calculated with the following formula (Figure 12.1):

$$F_x = \frac{\rho}{2} \cdot C_x \cdot S \cdot V^2$$

Where:
ρ = density of air (1.2/1.3 kg/m³)
C_x = aerodynamic drag coefficient
S = cross-section of the body (perpendicular to the airflow) (m²)
V = wind speed (m/s)

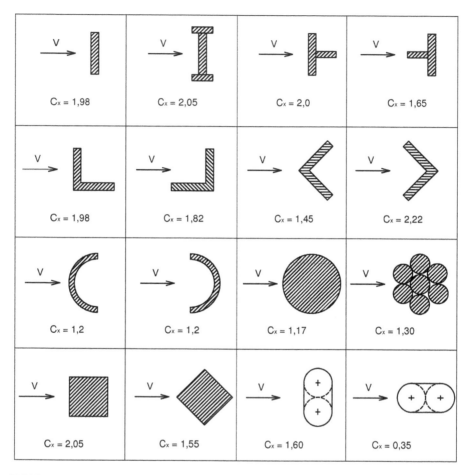

FIGURE 12.1
Aerodynamic drag coefficients C_x for different profiles.

12.3 Aerodynamic Characteristics of Blunt and Streamlined Bodies

This section presents the aerodynamic drag coefficients, C_x, corresponding to different shapes of hubs and nacelles, useful for the calculation of the aerodynamic loads on support structures. The C_x of blunt and streamlined bodies varies with Re, hence, the most complete data are always provided as curves. In this section, only average coefficients are provided, valid in turbulent flow conditions (Re > 600,000). Remember that a body is considered "streamlined" when its length is at least three times its diameter. The ratio L/D is called *aspect ratio* and it influences the drag. In general,

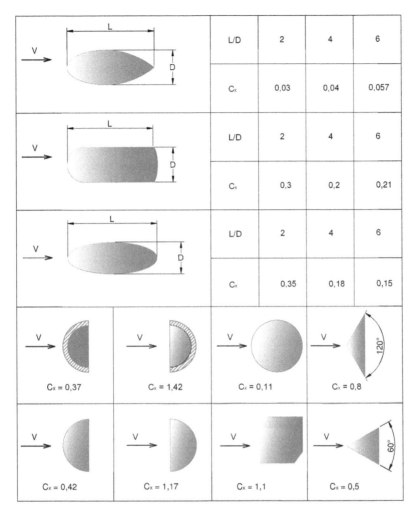

FIGURE 12.2
Drag coefficients, C_x, for different shapes of nacelles ad hubs, valid for Re > 600.000.

streamlined bodies having $L/D \approx 4$ present the minimum drag possible (Figure 12.2).

12.4 Aerodynamic Characteristics of Airfoils

This section contains the characteristics of some airfoils that are particularly interesting for the construction of small size wind turbines. Data are presented as tables instead of polar curves, in order to make reading and interpolation easier. The airfoil geometric data are in *Lednicer format*, i.e., as a table with two separate columns, each one corresponding, respectively, to the coordinates of the intrados and extrados. The convention of the *Lednicer* format, adopted by most programs for aerodynamic simulation, states that $x = 0$ corresponds to the leading edge, $x = 1$ corresponds to the trailing edge, and points of the extrados are always in the positive quadrant, while those of the intrados can be positive or negative, depending on the airfoil's shape. The aerodynamic data presented in the current edition were generated with Xfoil©, a program developed by Mike Drela and Harold Youngren of MIT in the 1980s and distributed freely under GNU License (http://web.mit.edu/drela/Public/web/xfoil/). Xfoil© calculates the lift and drag of any shape as a function of Re, given values very coincident with those measured in the wind tunnel, on condition that the attack angle is smaller than the stall angle. The Author has calculated the aerodynamic coefficients in two conditions that are common in the design of small wind turbines: Re = 200,000 and Re = 500,000, using the online version of Xfoil© accessible from www.airfoiltools.com. The data are rounded to just three decimals because, considering the influence of Re, the rugosity of the airfoil's surface, and the manufacture imperfections on the aerodynamic performance, higher accuracy is not justified. Please remember that the performances of airfoils become almost independent of Re, when the latter is bigger than 600,000. For very low Re values (smaller than 200,000), corresponding to the design of turbines able to run on very weak winds, or in the case of very tiny turbines ($D < 2\,\mathrm{m}$), it is advised to employ specific aerodynamic data for the resulting Re. Such data are available in the aircraft modelling manuals cited in the Bibliography. Alternatively, it is possible to generate a suitable set of data with Xfoil©, stating the specific Re of the case.

12.4.1 Clark Y Airfoil

This is a flat—convex airfoil, developed by *Col. Virginius Clark* during World War II. Its aerodynamic performance is mediocre, but since it is relatively easy to build with plywood and simple tools, the wings of several military planes of the time were built with such airfoil. It is highly appreciated by amateur aircraft modelers because of it construction simplicity. Its relative thickness is 11.7%, presenting a good compromise between rigidity, structural resistance, and aerodynamic efficiency (Figure 12.3; Tables 12.1 and 12.2).

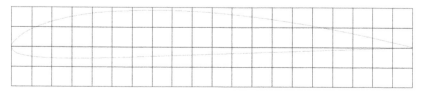

FIGURE 12.3
Geometry of the Clark Y 11.7% airfoil.

TABLE 12.1

Coordinates of the Clark Y 11.7% Airfoil

Extrados		Intrados	
x	y	x	y
0	0.00047	0	0.00047
0.00107	0.00616	0.00107	−0.00453
0.00428	0.01254	0.00428	−0.00898
0.00961	0.01943	0.00961	−0.01296
0.01704	0.02652	0.01704	−0.01651
0.02653	0.03352	0.02653	−0.01959
0.03806	0.04027	0.03806	−0.02214
0.05156	0.04677	0.05156	−0.02414
0.06699	0.05313	0.06699	−0.02567
0.08427	0.05939	0.08427	−0.0268
0.10332	0.06552	0.10332	−0.02763
0.12408	0.07134	0.12408	−0.02816
0.14645	0.0766	0.14645	−0.02839
0.17033	0.08113	0.17033	−0.02832
0.19562	0.08483	0.19562	−0.02795
0.22221	0.08774	0.22221	−0.02734
0.25	0.08996	0.25	−0.02653
0.27886	0.09158	0.27886	−0.02559
0.30866	0.09266	0.30866	−0.02458
0.33928	0.09318	0.33928	−0.02351
0.37059	0.09312	0.37059	−0.02242
0.43474	0.09128	0.43474	−0.02018
0.5	0.08719	0.5	−0.01792
0.56526	0.08105	0.56526	−0.01566
0.62941	0.07319	0.62941	−0.01345
0.69134	0.06405	0.69134	−0.01131
0.75	0.05412	0.75	−0.00928
0.80438	0.04394	0.80438	−0.00741
0.85355	0.034	0.85355	−0.00575
0.89668	0.02475	0.89668	−0.00429
0.93301	0.01656	0.93301	−0.00302
0.96194	0.00972	0.96194	−0.0019
0.98296	0.00448	0.98296	−0.00094
0.99572	0.00115	0.99572	−0.00025
1	0	1	0

TABLE 12.2

Aerodynamic Coefficients of Clark Y 11.5% Airfoil

	Re = 200,000				Re = 500,000			
α	C_z	C_x	C_m	f	C_z	C_x	C_m	f
−8.5	−0.418	0.090	−0.038	−4.63	−0.529	0.020	−0.086	−26.07
−8	−0.470	0.081	−0.047	−5.81	−0.464	0.018	−0.089	−25.95
−7.5	−0.435	0.073	−0.048	−5.95	−0.405	0.016	−0.091	−24.66
−7	−0.452	0.041	−0.068	−11.11	−0.340	0.015	−0.093	−23.08
−6.5	−0.415	0.031	−0.069	−13.41	−0.281	0.014	−0.093	−20.25
−6	−0.361	0.026	−0.071	−13.95	−0.221	0.013	−0.095	−17.40
−5.5	−0.304	0.024	−0.071	−12.77	−0.165	0.012	−0.094	−13.99
−5	−0.229	0.021	−0.075	−10.77	−0.113	0.011	−0.093	−10.16
−4.5	−0.170	0.020	−0.076	−8.70	−0.060	0.011	−0.093	−5.70
−4	−0.087	0.018	−0.081	−4.87	−0.008	0.010	−0.092	−0.78
−3.5	−0.027	0.017	−0.081	−1.63	0.044	0.010	−0.091	4.56
−3	0.036	0.015	−0.083	2.33	0.096	0.009	−0.090	10.30
−2.5	0.107	0.014	−0.085	7.64	0.146	0.009	−0.088	16.47
−2	0.160	0.013	−0.085	12.37	0.194	0.008	−0.086	23.06
−1.5	0.204	0.012	−0.082	17.60	0.243	0.008	−0.085	30.30
−1	0.340	0.011	−0.095	32.00	0.289	0.007	−0.083	38.60
−0.5	0.477	0.010	−0.111	45.60	0.325	0.007	−0.078	48.70
0	0.523	0.010	−0.109	50.62	0.425	0.007	−0.087	65.15
0.5	0.564	0.010	−0.106	54.51	0.513	0.007	−0.093	75.03
1	0.611	0.010	−0.104	58.87	0.607	0.007	−0.101	85.90
1.5	0.653	0.010	−0.101	62.35	0.669	0.007	−0.103	93.09
2	0.698	0.011	−0.098	65.77	0.711	0.007	−0.100	96.42
2.5	0.742	0.011	−0.096	68.84	0.752	0.008	−0.097	99.05
3	0.785	0.011	−0.092	71.19	0.793	0.008	−0.094	100.70
3.5	0.827	0.011	−0.089	72.95	0.833	0.008	−0.090	101.56
4	0.866	0.012	−0.086	74.04	0.871	0.009	−0.087	100.96
4.5	0.906	0.012	−0.082	74.36	0.908	0.009	−0.083	98.99
5	0.944	0.013	−0.079	73.95	0.945	0.010	−0.079	97.10
5.5	0.980	0.013	−0.075	72.95	0.983	0.010	−0.076	95.72
6	1.015	0.014	−0.071	71.76	1.021	0.011	−0.072	94.28
6.5	1.052	0.015	−0.067	70.75	1.0604	0.011	−0.069	93.59
7	1.089	0.016	−0.064	69.93	1.0992	0.012	−0.066	92.84
7.5	1.126	0.016	−0.060	68.97	1.1368	0.012	−0.063	91.83
8	1.161	0.017	−0.057	68.03	1.1704	0.013	−0.058	90.24
8.5	1.193	0.018	−0.052	66.84	1.2037	0.014	−0.054	88.83
9	1.217	0.019	−0.047	65.42	1.2308	0.014	−0.049	84.94
9.5	1.237	0.020	−0.041	63.35	1.2406	0.016	−0.042	75.37
10	1.246	0.021	−0.034	59.13	1.261	0.018	−0.037	69.94

(Continued)

TABLE 12.2 (*Continued*)

Aerodynamic Coefficients of Clark Y 11.5% Airfoil

	Re = 200,000				Re = 500,000			
α	C_z	C_x	C_m	f	C_z	C_x	C_m	f
10.5	1.245	0.024	−0.027	52.29	1.2847	0.019	−0.033	65.92
11	1.252	0.027	−0.022	47.16	1.3043	0.021	−0.028	61.29
11.5	1.262	0.029	−0.017	42.91	1.3295	0.023	−0.025	58.21
12	1.272	0.032	−0.013	39.17	1.345	0.025	−0.021	53.48
12.5	1.287	0.035	−0.010	36.28	1.3697	0.027	−0.019	50.79
13	1.303	0.039	−0.008	33.80	1.3772	0.030	−0.015	45.62
13.5	1.319	0.042	−0.005	31.59	1.4006	0.032	−0.013	43.24
14	1.331	0.045	−0.003	29.37	1.4126	0.036	−0.011	39.61
14.5	1.344	0.049	−0.001	27.44	1.4214	0.039	−0.009	36.03
15	1.349	0.053	0.000	25.23	1.434	0.043	−0.008	33.26
15.5	1.352	0.058	0.001	23.30	1.4327	0.048	−0.007	29.56
16	1.353	0.064	0.002	21.29	1.4404	0.053	−0.007	27.10
16.5	1.346	0.070	0.001	19.27	1.4283	0.060	−0.008	23.63
17	1.348	0.076	0.001	17.85	1.4316	0.066	−0.009	21.64
17.5	1.329	0.084	−0.002	15.75	1.4185	0.074	−0.010	19.09
18	1.315	0.093	−0.005	14.14	1.4032	0.083	−0.013	16.91
18.5	1.314	0.100	−0.006	13.20	1.3895	0.092	−0.015	15.15
19	1.278	0.113	−0.013	11.33	1.3682	0.102	−0.019	13.43

N.B.: the pitching moment is referred to $x = 0.25$; $y = 0$; $y = 0$.

12.4.2 Wortmann FX77-W153 Airfoil

This is a biconvex airfoil, featuring high aerodynamic performance, specially conceived for the construction of wind turbines, in such a way that its lift and drag remain almost independent of the rugosity. The relative thickness is 15.2%, so it provides robust blade sections, enough for the structural safety of small rotors (Figure 12.4; Tables 12.3 and 12.4).

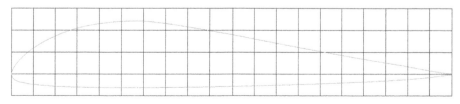

FIGURE 12.4
Geometry of the FX77-W153 airfoil.

TABLE 12.3

Coordinates of the FX77-W153 Airfoil

Extrados		Intrados	
x	*y*	*x*	*y*
0	0	0	0
0.00107	0.00591	0.00107	−0.00464
0.00428	0.01365	0.00428	−0.01115
0.00961	0.02143	0.00961	−0.01482
0.01704	0.03089	0.01704	−0.01817
0.02653	0.04032	0.02653	−0.02016
0.03806	0.05048	0.03806	−0.02255
0.05156	0.06014	0.05156	−0.02414
0.06699	0.07013	0.06699	−0.02598
0.08427	0.07928	0.08427	−0.02713
0.10332	0.08836	0.10332	−0.02849
0.12408	0.09639	0.12408	−0.02929
0.14645	0.10372	0.14645	−0.03021
0.17033	0.10954	0.17033	−0.03064
0.19562	0.11449	0.19562	−0.03123
0.22221	0.11791	0.22221	−0.03134
0.25	0.12037	0.25	−0.03166
0.27886	0.12105	0.27886	−0.03143
0.30866	0.1204	0.30866	−0.03154
0.33928	0.11696	0.33928	−0.03105
0.37059	0.1128	0.37059	−0.03088
0.40245	0.10758	0.40245	−0.03025
0.43474	0.10209	0.43474	−0.02984
0.4673	0.09606	0.4673	−0.02908
0.5	0.08997	0.5	−0.02846
0.5327	0.08363	0.5327	−0.02757
0.56526	0.0774	0.56526	−0.02681
0.59755	0.07113	0.59755	−0.02578
0.62941	0.06512	0.62941	−0.0249
0.66072	0.05913	0.66072	−0.02376
0.69134	0.05355	0.69134	−0.02276
0.72114	0.04802	0.72114	−0.02152
0.75	0.04291	0.75	−0.02044
0.77779	0.03794	0.77779	−0.01914
0.80438	0.03339	0.80438	−0.01797
0.82967	0.02902	0.82967	−0.01662
0.85355	0.02503	0.85355	−0.01536
0.87592	0.02118	0.87592	−0.01397
0.89668	0.01758	0.89668	−0.01263
0.91573	0.01408	0.91573	−0.01115
0.93301	0.0109	0.93301	−0.00962

(Continued)

TABLE 12.3 (*Continued*)

Coordinates of the FX77-W153 Airfoil

	Extrados		Intrados
x	*y*	*x*	*y*
0.94844	0.00801	0.94844	−0.00793
0.96194	0.00561	0.96194	−0.00627
0.97347	0.00367	0.97347	−0.00464
0.98296	0.00231	0.98296	−0.00325
0.99039	0.00143	0.99039	−0.00216
0.99572	0.00099	0.99572	−0.00141
0.99893	0.00084	0.99893	−0.00097

TABLE 12.4

Aerodynamic Coefficients of FX77-W153 Airfoil

	Re = 200,000				Re = 500,000			
α	C_z	C_x	C_m	*f*	C_z	C_x	C_m	*f*
−8.5	−0.339	0.095	−0.008	−3.56	−0.395	0.079	−0.019	−5.01
−8	−0.369	0.082	−0.027	−4.50	−0.404	0.069	−0.028	−5.88
−7.5	−0.364	0.073	−0.030	−4.96	−0.398	0.060	−0.032	−6.58
−7	−0.342	0.068	−0.030	−5.01	−0.383	0.053	−0.034	−7.25
−6.5	−0.343	0.058	−0.035	−5.92	−0.363	0.045	−0.034	−8.05
−6	−0.308	0.053	−0.035	−5.80	−0.368	0.026	−0.028	−14.09
−5.5	−0.282	0.047	−0.035	−6.02	−0.326	0.022	−0.025	−15.05
−5	−0.242	0.043	−0.034	−5.69	−0.277	0.019	−0.024	−14.69
−4.5	−0.209	0.028	−0.028	−7.38	−0.225	0.017	−0.022	−13.06
−4	−0.159	0.025	−0.026	−6.38	−0.171	0.016	−0.022	−10.62
−3.5	−0.105	0.022	−0.025	−4.67	−0.118	0.015	−0.021	−7.87
−3	−0.050	0.021	−0.025	−2.38	−0.066	0.014	−0.020	−4.64
−2.5	0.003	0.020	−0.024	0.16	−0.012	0.014	−0.019	−0.89
−2	0.057	0.019	−0.023	2.97	0.043	0.013	−0.019	3.24
−1.5	0.110	0.019	−0.023	5.89	0.099	0.013	−0.019	7.41
−1	0.162	0.018	−0.022	8.93	0.154	0.013	−0.018	11.78
−0.5	0.257	0.015	−0.029	16.65	0.210	0.013	−0.018	16.28
0	0.393	0.016	−0.044	23.95	0.263	0.012	−0.018	21.52
0.5	0.448	0.017	−0.045	26.62	0.306	0.011	−0.016	28.11
1	0.498	0.017	−0.044	28.98	0.401	0.010	−0.023	38.61
1.5	0.548	0.018	−0.043	31.12	0.484	0.011	−0.028	44.75
2	0.598	0.018	−0.043	32.60	0.593	0.011	−0.039	53.19
2.5	0.647	0.019	−0.042	34.13	0.656	0.011	−0.041	58.32
3	0.697	0.019	−0.041	35.97	0.709	0.011	−0.041	62.07
3.5	0.746	0.020	−0.040	36.99	0.761	0.012	−0.040	65.69
4	0.792	0.021	−0.039	37.40	0.814	0.012	−0.039	68.85

(*Continued*)

TABLE 12.4 (*Continued*)

Aerodynamic Coefficients of FX77-W153 Airfoil

	Re = 200,000				Re = 500,000			
α	C_z	C_x	C_m	f	C_z	C_x	C_m	f
4.5	0.840	0.022	−0.038	38.67	0.866	0.012	−0.039	70.52
5	0.891	0.022	−0.037	40.13	0.918	0.012	−0.038	73.83
5.5	0.932	0.024	−0.036	39.65	0.969	0.013	−0.038	76.99
6	0.980	0.024	−0.034	40.82	1.021	0.013	−0.037	79.88
6.5	1.032	0.024	−0.034	42.38	1.072	0.013	−0.037	81.17
7	1.067	0.026	−0.032	41.11	1.122	0.013	−0.036	83.56
7.5	1.115	0.026	−0.031	42.32	1.172	0.014	−0.035	86.20
8	1.166	0.027	−0.030	43.43	1.222	0.014	−0.034	88.71
8.5	1.195	0.029	−0.027	41.87	1.271	0.014	−0.033	90.89
9	1.247	0.029	−0.026	43.63	1.319	0.014	−0.032	94.50
9.5	1.274	0.031	−0.023	41.76	1.367	0.014	−0.031	97.22
10	1.318	0.031	−0.022	42.74	1.415	0.014	−0.030	100.26
10.5	1.380	0.030	−0.022	46.08	1.461	0.014	−0.029	101.82
11	1.402	0.031	−0.019	45.20	1.502	0.015	−0.027	101.51
11.5	1.462	0.030	−0.019	49.04	1.540	0.016	−0.025	98.77
12	1.475	0.031	−0.014	47.16	1.576	0.016	−0.023	95.83
12.5	1.507	0.031	−0.012	48.14	1.603	0.018	−0.019	90.82
13	1.538	0.031	−0.009	49.85	1.618	0.019	−0.015	84.03
13.5	1.534	0.034	−0.005	45.55	1.631	0.021	−0.011	76.34
14	1.568	0.034	−0.004	45.64	1.633	0.025	−0.009	64.97
14.5	1.550	0.040	−0.004	38.46	1.614	0.033	−0.011	49.39
15	1.501	0.051	−0.006	29.47	1.583	0.042	−0.012	37.72
15.5	1.500	0.056	−0.007	26.69	1.547	0.052	−0.013	30.03
16	1.510	0.060	−0.007	25.20	1.499	0.062	−0.015	24.08
16.5	1.417	0.077	−0.012	18.29	1.452	0.074	−0.017	19.67
17	1.280	0.103	−0.019	12.41	1.416	0.084	−0.019	16.77
17.5	1.364	0.097	−0.017	14.10				

N.B.: the pitching moment is referred to $x = 0.25$; $y = 0$.

12.4.3 Eppler E220 Airfoil

This is a biconvex airfoil, featuring high aerodynamic performance, specially conceived for the construction of wind turbines, in such a way that its lift and drag remain almost independent of the rugosity. The relative thickness is 11.48% (Figure 12.5; Tables 12.5 and 12.6).

12.4.4 Airfoil NREL S822

This is a biconvex airfoil, featuring high aerodynamic performance, specially conceived for the construction of wind turbines and hence, its lift and

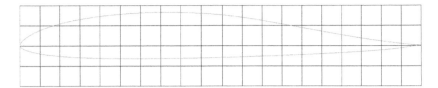

FIGURE 12.5
Geometry of the Eppler E220 airfoil.

TABLE 12.5

Coordinates of Eppler E220 Airfoil

Extrados		Intrados	
x	y	x	y
0.00148	0.00452	0.00148	0.00452
0.00761	0.01241	0.00025	−0.00172
0.01818	0.02095	0.00492	−0.0069
0.03307	0.02968	0.01552	−0.01208
0.0522	0.0383	0.03134	−0.01689
0.07543	0.04658	0.05221	−0.02113
0.10256	0.05432	0.07795	−0.02474
0.13337	0.06136	0.10834	−0.02769
0.16757	0.06754	0.14312	−0.02996
0.20485	0.07275	0.18194	−0.03161
0.24486	0.07687	0.2244	−0.03266
0.2872	0.07983	0.27004	−0.03316
0.33144	0.08152	0.31838	−0.03318
0.37714	0.08188	0.36885	−0.03277
0.42381	0.08072	0.42089	−0.03199
0.47113	0.07784	0.4739	−0.03088
0.51893	0.07317	0.52725	−0.02949
0.56707	0.06675	0.58032	−0.02787
0.61555	0.0589	0.6325	−0.02604
0.66408	0.05031	0.68318	−0.02404
0.71202	0.04156	0.73176	−0.02191
0.75865	0.03311	0.77767	−0.01967
0.80322	0.02531	0.82037	−0.01733
0.84494	0.01844	0.85936	−0.01492
0.88303	0.01269	0.89416	−0.01244
0.91671	0.00809	0.92435	−0.00981
0.94537	0.00456	0.94977	−0.00691
0.96855	0.00206	0.97051	−0.00397
0.98575	0.00062	0.98633	−0.00162
0.99639	0.00008	0.99647	−0.00034
1	0	1	0

TABLE 12.6

Aerodynamic Coefficients of Eppler E220 Airfoil

	Re = 200,000				Re = 500,000			
α	C_z	C_x	C_m	f	C_z	C_x	C_m	f
−10.50	−0.351	0.105	−0.018	−3.33				
−10.00	−0.450	0.103	−0.017	−4.37				
−9.50	−0.441	0.096	−0.021	−4.61	−0.397	0.084	−0.017	−4.75
−9.00	−0.442	0.087	−0.029	−5.11	−0.408	0.073	−0.022	−5.62
−8.50	−0.380	0.067	−0.038	−5.70	−0.435	0.060	−0.030	−7.30
−8.00	−0.407	0.057	−0.044	−7.11	−0.492	0.046	−0.036	−10.69
−7.50	−0.428	0.049	−0.044	−8.68	−0.504	0.041	−0.036	−12.36
−7.00	−0.465	0.052	−0.046	−8.98	−0.511	0.034	−0.031	−15.01
−6.50	−0.456	0.045	−0.043	−10.23	−0.517	0.027	−0.026	−19.13
−6.00	−0.439	0.029	−0.036	−15.11	−0.483	0.019	−0.022	−25.31
−5.50	−0.399	0.024	−0.032	−16.58	−0.436	0.016	−0.020	−26.90
−5.00	−0.350	0.021	−0.031	−16.53	−0.386	0.015	−0.019	−26.37
−4.50	−0.300	0.018	−0.029	−16.35	−0.336	0.013	−0.018	−24.96
−4.00	−0.253	0.017	−0.027	−15.21	−0.287	0.012	−0.017	−23.47
−3.50	−0.208	0.015	−0.025	−13.98	−0.237	0.011	−0.015	−21.20
−3.00	−0.169	0.013	−0.023	−12.83	−0.187	0.010	−0.014	−18.27
−2.50	−0.136	0.012	−0.018	−11.64	−0.138	0.009	−0.013	−14.77
−2.00	−0.005	0.012	−0.031	−0.42	−0.092	0.008	−0.012	−10.85
−1.50	0.094	0.013	−0.039	7.46	−0.055	0.007	−0.008	−7.61
−1.00	0.201	0.013	−0.049	15.89	0.054	0.007	−0.018	7.43
−0.50	0.273	0.013	−0.054	21.85	0.123	0.008	−0.020	16.00
0.00	0.319	0.013	−0.052	25.37	0.205	0.008	−0.026	25.78
0.25	0.343	0.013	−0.052	27.14	0.250	0.008	−0.029	31.07
0.50	0.366	0.013	−0.051	28.94	0.289	0.008	−0.032	35.44
0.75	0.390	0.013	−0.051	30.67	0.334	0.008	−0.035	41.22
1.00	0.414	0.013	−0.050	32.35	0.375	0.008	−0.038	46.44
1.50	0.462	0.013	−0.049	35.81	0.457	0.008	−0.044	58.01
2.00	0.511	0.013	−0.048	39.28	0.506	0.008	−0.043	64.40
2.50	0.559	0.013	−0.046	42.59	0.557	0.008	−0.042	71.16
3.00	0.608	0.013	−0.045	45.93	0.607	0.008	−0.041	77.63
3.50	0.656	0.013	−0.044	49.49	0.658	0.008	−0.041	84.13
4.00	0.705	0.013	−0.042	53.30	0.709	0.008	−0.040	90.54
4.50	0.754	0.013	−0.040	57.42	0.760	0.008	−0.039	96.53
5.00	0.803	0.013	−0.039	62.10	0.810	0.008	−0.038	101.94
5.50	0.851	0.013	−0.037	66.70	0.860	0.008	−0.037	105.39
6.00	0.899	0.013	−0.035	71.39	0.907	0.009	−0.036	104.76
6.50	0.945	0.013	−0.033	75.15	0.949	0.010	−0.034	97.54
7.00	0.986	0.013	−0.030	75.86	0.987	0.011	−0.032	89.05
7.50	1.018	0.014	−0.027	71.29	1.022	0.012	−0.029	82.30

(Continued)

TABLE 12.6 (*Continued*)

Aerodynamic Coefficients of Eppler E220 Airfoil

	Re = 200,000				Re = 500,000			
α	C_z	C_x	C_m	f	C_z	C_x	C_m	f
8.00	1.039	0.016	−0.022	64.02	1.055	0.014	−0.026	76.79
8.50	1.055	0.018	−0.016	57.81	1.084	0.015	−0.022	71.70
9.00	1.064	0.020	−0.010	52.17	1.107	0.017	−0.018	66.65
9.50	1.057	0.023	−0.002	46.47	1.120	0.018	−0.011	60.99
10.00	1.051	0.026	0.005	40.56	1.115	0.020	−0.002	55.26
10.50	1.053	0.029	0.009	36.02	1.127	0.022	0.002	51.14
11.00	1.057	0.033	0.013	32.08	1.129	0.025	0.007	44.86
11.50	1.068	0.036	0.016	29.46	1.143	0.028	0.010	41.16
12.00	1.086	0.040	0.019	27.36	1.153	0.031	0.013	37.42
12.50	1.101	0.043	0.022	25.55	1.150	0.035	0.016	32.57
13.00	1.118	0.048	0.024	23.29	1.161	0.039	0.017	29.97
13.50	1.122	0.053	0.026	21.05	1.165	0.043	0.019	27.20
14.00	1.113	0.060	0.027	18.54	1.165	0.048	0.020	24.42
14.50	1.097	0.067	0.027	16.35	1.154	0.054	0.022	21.35
15.00	1.076	0.076	0.026	14.25	1.154	0.060	0.022	19.37
15.50	1.053	0.085	0.023	12.43	1.149	0.066	0.021	17.35
16.00	0.983	0.104	0.013	9.41	1.140	0.074	0.019	15.45
16.50	0.923	0.126	0.000	7.32	1.124	0.083	0.017	13.54
17.00					1.108	0.093	0.013	11.92
17.50					1.089	0.104	0.008	10.48
18.00					1.052	0.119	0.000	8.84

Blank cells indicate that the airfoil is stalled and hence, it is impossible to calculate a reliable value.
N.B.: the pitching moment is referred to $x = 0.25; y = 0$.

drag coefficients are almost independent of the rugosity. The relative thickness is 16%, which is robust enough for building rotors up to 5 m in diameter (Figure 12.6; Tables 12.7 and 12.8).

12.4.5 Airfoil NREL S819

This is a biconvex airfoil, featuring high aerodynamic performance, specially conceived for the construction of wind turbines and hence, its lift and

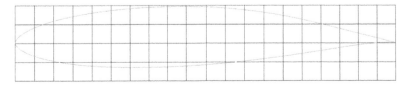

FIGURE 12.6
Geometry of the NREL S822 airfoil.

TABLE 12.7

Coordinates of the Airfoil NREL S822

Extrados		Intrados	
x	y	x	y
0.000138	0.001358	0.000138	0.001358
0.000651	0.003396	0.000023	−0.000542
0.00176	0.006074	0.000294	−0.001935
0.008599	0.015142	0.000443	−0.002413
0.02031	0.024706	0.001282	−0.004385
0.036592	0.034302	0.005329	−0.010118
0.057423	0.043713	0.015376	−0.01868
0.082528	0.052754	0.030216	−0.027173
0.111706	0.061261	0.049559	−0.035201
0.144635	0.069084	0.073374	−0.042563
0.181021	0.076101	0.101374	−0.049059
0.220471	0.082211	0.133441	−0.054581
0.262594	0.087322	0.169219	−0.059031
0.306932	0.091363	0.208482	−0.062361
0.353012	0.094259	0.250797	−0.064545
0.400324	0.095949	0.295834	−0.065585
0.44832	0.096342	0.343087	−0.065504
0.496467	0.095295	0.392138	−0.06434
0.544321	0.092477	0.44242	−0.062144
0.591971	0.087923	0.493444	−0.058965
0.638946	0.081988	0.544594	−0.054833
0.684771	0.074819	0.595407	−0.049596
0.728946	0.066315	0.645795	−0.043462
0.771742	0.056677	0.695149	−0.036874
0.812782	0.046731	0.742856	−0.030066
0.851444	0.037037	0.788279	−0.023032
0.887037	0.028045	0.831428	−0.016082
0.918825	0.020084	0.871626	−0.010003
0.946046	0.013324	0.907952	−0.005223
0.968133	0.007602	0.93944	−0.001945
0.985048	0.003157	0.96515	−0.000135
0.996089	0.000642	0.984246	0.000448
1	0	0.996022	0.000259
		1	0

TABLE 12.8

Aerodynamic Coefficients of NREL S822 Airfoil

	Re = 200,000				Re = 500,000			
α	C_z	C_x	C_m	f	C_z	C_x	C_m	f
−17.50					−0.797	0.091	−0.072	−8.72
−17.00					−0.827	0.077	−0.079	−10.80
−16.50	−0.561	0.121	−0.062	−4.64	−0.844	0.071	−0.081	−11.89
−16.00	−0.613	0.105	−0.070	−5.82	−0.854	0.067	−0.083	−12.83
−15.50	−0.694	0.083	−0.083	−8.37	−0.884	0.054	−0.087	−16.39
−15.00	−0.746	0.069	−0.090	−10.81	−0.896	0.047	−0.088	−18.95
−14.50	−0.779	0.060	−0.092	−12.98	−0.905	0.042	−0.088	−21.49
−14.00	−0.808	0.053	−0.092	−15.19	−0.917	0.038	−0.085	−24.27
−13.50	−0.836	0.048	−0.088	−17.31	−0.907	0.034	−0.087	−27.05
−13.00	−0.881	0.046	−0.079	−19.22	−0.878	0.029	−0.092	−30.54
−12.50	−0.916	0.043	−0.070	−21.11	−0.838	0.026	−0.096	−32.74
−12.00	−0.892	0.038	−0.073	−23.37	−0.791	0.023	−0.100	−34.69
−11.50	−0.862	0.035	−0.075	−24.44	−0.745	0.020	−0.103	−36.72
−11.00	−0.831	0.032	−0.075	−25.97	−0.707	0.018	−0.104	−38.46
−10.50	−0.794	0.029	−0.076	−27.15	−0.679	0.017	−0.101	−39.70
−10.00	−0.770	0.027	−0.074	−28.19	−0.654	0.016	−0.098	−40.82
−9.50	−0.743	0.026	−0.072	−28.95	−0.630	0.015	−0.093	−41.86
−9.00	−0.705	0.024	−0.071	−29.40	−0.604	0.014	−0.089	−42.09
−8.50	−0.683	0.023	−0.067	−30.01	−0.569	0.014	−0.086	−41.92
−8.00	−0.629	0.021	−0.068	−29.53	−0.531	0.013	−0.083	−41.14
−7.50	−0.608	0.021	−0.063	−29.55	−0.490	0.012	−0.082	−39.98
−7.00	−0.546	0.019	−0.066	−28.10	−0.447	0.012	−0.080	−38.21
−6.50	−0.526	0.019	−0.060	−27.62	−0.403	0.011	−0.078	−36.18
−6.00	−0.457	0.018	−0.064	−25.20	−0.356	0.011	−0.078	−33.42
−5.50	−0.437	0.018	−0.058	−24.30	−0.310	0.010	−0.076	−30.43
−5.00	−0.366	0.017	−0.062	−21.14	−0.261	0.010	−0.076	−26.72
−4.50	−0.343	0.017	−0.056	−19.76	−0.211	0.009	−0.075	−22.41
−4.00	−0.271	0.017	−0.059	−15.91	−0.160	0.009	−0.074	−17.68
−3.50	−0.244	0.017	−0.054	−14.02	−0.107	0.009	−0.074	−12.19
−3.00	−0.172	0.017	−0.057	−9.90	−0.056	0.009	−0.074	−6.47
−2.50	−0.145	0.018	−0.052	−8.05	0.000	0.009	−0.074	−0.02
−2.00	−0.073	0.018	−0.054	−4.07	0.055	0.008	−0.074	6.47
−1.50	−0.048	0.019	−0.049	−2.55	0.110	0.008	−0.074	13.02
−1.00	−0.014	0.019	−0.045	−0.72	0.168	0.008	−0.074	19.89
−0.50	0.045	0.020	−0.045	2.26	0.222	0.008	−0.074	26.22
0.00	0.118	0.020	−0.047	5.97	0.280	0.008	−0.075	33.12
0.50	0.140	0.020	−0.042	6.84	0.334	0.009	−0.075	39.22
1.00	0.216	0.020	−0.044	10.70	0.393	0.008	−0.075	46.34
1.50	0.238	0.021	−0.038	11.47	0.446	0.008	−0.075	53.07

(Continued)

TABLE 12.8 (*Continued*)

Aerodynamic Coefficients of NREL S822 Airfoil

	Re = 200,000				Re = 500,000			
α	C_z	C_x	C_m	f	C_z	C_x	C_m	f
2.00	0.317	0.020	−0.041	15.75	0.500	0.008	−0.074	59.64
2.50	0.305	0.021	−0.029	14.40	0.554	0.008	−0.074	66.29
3.00	0.379	0.020	−0.032	18.74	0.608	0.008	−0.073	72.73
3.50	0.474	0.019	−0.037	25.08	0.660	0.008	−0.073	78.74
4.00	0.525	0.018	−0.036	28.58	0.710	0.008	−0.071	84.19
4.50	0.598	0.017	−0.038	34.19	0.758	0.009	−0.070	89.05
5.00	0.683	0.016	−0.043	41.86	0.802	0.009	−0.067	92.49
5.50	0.779	0.015	−0.050	51.47	0.842	0.009	−0.064	94.57
6.00	0.857	0.014	−0.054	59.51	0.869	0.009	−0.059	93.57
6.50	0.928	0.014	−0.057	65.71	0.865	0.010	−0.047	85.09
7.00	0.987	0.014	−0.058	68.56	0.863	0.011	−0.037	75.53
7.50	0.949	0.015	−0.042	64.32	0.867	0.013	−0.028	68.61
8.00	0.926	0.016	−0.029	58.04	0.900	0.014	−0.026	63.36
8.50	0.927	0.018	−0.022	51.37	0.930	0.016	−0.024	58.74
9.00	0.934	0.020	−0.017	45.76	0.958	0.017	−0.021	55.07
9.50	0.945	0.023	−0.012	41.30	0.987	0.019	−0.019	51.95
10.00	0.959	0.025	−0.008	37.71	1.015	0.021	−0.016	49.35
10.50	0.977	0.028	−0.005	34.89	1.042	0.022	−0.014	46.68
11.00	0.996	0.031	−0.002	32.65	1.067	0.024	−0.011	44.07
11.50	1.015	0.033	0.001	30.46	1.092	0.026	−0.008	41.73
12.00	1.035	0.036	0.003	28.64	1.112	0.029	−0.006	38.98
12.50	1.058	0.039	0.005	27.02	1.133	0.031	−0.003	36.59
13.00	1.085	0.042	0.007	25.58	1.151	0.034	−0.001	34.15
13.50	1.106	0.046	0.009	24.30	1.172	0.036	0.001	32.25
14.00	1.124	0.049	0.010	22.95	1.177	0.040	0.003	29.16
14.50	1.141	0.054	0.012	21.26	1.196	0.044	0.005	27.49
15.00	1.146	0.059	0.013	19.57	1.213	0.047	0.005	25.86
15.50	1.150	0.063	0.013	18.14	1.209	0.053	0.007	22.96
16.00	1.139	0.072	0.013	15.88	1.221	0.057	0.007	21.39
16.50	1.114	0.081	0.012	13.77	1.229	0.062	0.006	19.78
17.00	1.081	0.092	0.008	11.70	1.222	0.069	0.005	17.63
17.50	1.038	0.107	0.002	9.72	1.215	0.077	0.003	15.76
18.00	0.985	0.125	−0.009	7.86	1.210	0.085	0.001	14.23
18.50	0.917	0.151	−0.025	6.09	1.201	0.094	−0.003	12.75
19.00					1.188	0.104	−0.008	11.40

N.B.: the pitching moment is referred to $x = 0.25$; $y = 0$.

drag coefficients are almost independent of the rugosity. The relative thickness is 21.1%, making it particularly robust and hence, suitable to implement the joint section between the blade and the hub in turbines having $D < 15\,\text{m}$ (Figure 12.7; Tables 12.9 and 12.10).

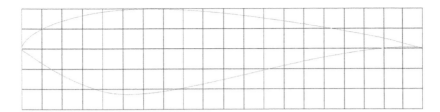

FIGURE 12.7

Geometry of the airfoil NREL S819.

TABLE 12.9

Coordinates of NREL S819 Airfoil

Extrados		Intrados	
x	*y*	*x*	*y*
0.001025	0.00543	0.001025	0.00543
0.006778	0.015626	0.000016	0.00063
0.017236	0.026575	0.000085	−0.001387
0.032137	0.037616	0.000273	−0.002303
0.051442	0.048461	0.000507	−0.002953
0.074857	0.058848	0.000592	−0.003128
0.102158	0.068539	0.001122	−0.00389
0.133009	0.077309	0.001853	−0.004672
0.16707	0.084912	0.003801	−0.006327
0.203951	0.091061	0.009034	−0.009898
0.243383	0.095223	0.025086	−0.020533
0.285654	0.097385	0.04469	−0.033671
0.330358	0.0978	0.067554	−0.047997
0.377132	0.096532	0.092912	−0.062604
0.425591	0.093712	0.120577	−0.076796
0.475305	0.089511	0.150045	−0.089856
0.525799	0.084135	0.181072	−0.101107
0.576554	0.077814	0.213187	−0.109658
0.62701	0.070795	0.246621	−0.113752
0.676568	0.063323	0.283625	−0.113579
0.724602	0.055635	0.323851	−0.110343
0.770468	0.047944	0.367148	−0.104334
0.813514	0.040423	0.413252	−0.096074
0.8531	0.033195	0.461795	−0.086042
0.888601	0.026289	0.512331	−0.07475
0.919582	0.019478	0.564325	−0.062742
0.946197	0.012848	0.617142	−0.050578
0.968321	0.007082	0.670041	−0.038808
0.985332	0.002884	0.72218	−0.027942

(Continued)

TABLE 12.9 (*Continued*)

Coordinates of NREL S819 Airfoil

Extrados		Intrados	
x	y	x	y
0.996227	0.00061	0.772632	−0.018422
1	0	0.820407	−0.010583
		0.864481	−0.004633
		0.903842	−0.000629
		0.937525	0.001531
		0.964668	0.002098
		0.984436	0.001434
		0.996165	0.000438
		1	0

TABLE 12.10

Aerodynamic Coefficients of NREL S819 Airfoil

	Re = 200,000				Re = 500,000			
α	C_z	C_x	C_m	f	C_z	C_x	C_m	f
					−0.640	0.162	−0.025	−4.0
					−0.676	0.146	−0.034	−4.6
					−0.709	0.132	−0.040	−5.4
					−0.740	0.121	−0.046	−6.1
					−0.768	0.110	−0.050	−7.0
					−0.793	0.101	−0.053	−7.8
−16.0	−0.865	0.095	−0.060	−9.1	−0.813	0.093	−0.056	−8.7
−15.5	−0.887	0.091	−0.062	−9.7	−0.830	0.086	−0.057	−9.7
−15.0	−0.927	0.085	−0.063	−10.9	−0.842	0.080	−0.058	−10.6
−14.5	−0.889	0.077	−0.062	−11.6	−0.849	0.074	−0.059	−11.5
−14.0	−0.894	0.072	−0.062	−12.5	−0.850	0.069	−0.058	−12.4
−13.5	−0.921	0.067	−0.061	−13.7	−0.844	0.064	−0.057	−13.1
−13.0	−0.970	0.065	−0.058	−14.9	−0.836	0.060	−0.056	−13.9
−12.5	−1.026	0.064	−0.050	−16.0	−0.839	0.057	−0.054	−14.8
−12.0	−0.844	0.053	−0.056	−16.0	−0.866	0.054	−0.052	−16.1
−11.5	−0.771	0.052	−0.052	−14.9	−0.756	0.052	−0.049	−14.6
−11.0	−0.738	0.050	−0.049	−14.7	−0.700	0.051	−0.046	−13.7
−10.5	−0.610	0.058	−0.041	−10.5	−0.684	0.049	−0.043	−14.0
−10.0	−0.544	0.060	−0.036	−9.0	−0.713	0.045	−0.042	−15.9
−9.5	−0.483	0.064	−0.031	−7.6	−0.579	0.049	−0.032	−11.8
−9.0	−0.463	0.061	−0.029	−7.6	−0.558	0.050	−0.029	−11.1

(Continued)

TABLE 12.10 (*Continued*)

Aerodynamic Coefficients of NREL S819 Airfoil

	Re = 200,000				Re = 500,000			
α	C_z	C_x	C_m	f	C_z	C_x	C_m	f
−8.5	−0.432	0.063	−0.023	−6.9	−0.555	0.047	−0.027	−11.8
−8.0	−0.419	0.061	−0.020	−6.9	−0.517	0.043	−0.031	−11.9
−7.5	−0.420	0.059	−0.016	−7.2	−0.478	0.040	−0.036	−12.0
−7.0	−0.455	0.054	−0.011	−8.5	−0.440	0.036	−0.040	−12.2
−6.5	−0.484	0.046	−0.014	−10.5	−0.402	0.033	−0.045	−12.4
−6.0	−0.487	0.033	−0.029	−14.6	−0.373	0.024	−0.055	−15.3
−5.5	−0.463	0.025	−0.041	−18.6	−0.369	0.018	−0.060	−20.8
−5.0	−0.400	0.022	−0.048	−18.6	−0.317	0.016	−0.063	−20.4
−4.5	−0.327	0.020	−0.053	−16.6	−0.270	0.014	−0.063	−19.4
−4.0	−0.273	0.016	−0.057	−17.3	−0.197	0.013	−0.063	−15.1
−3.5	−0.198	0.017	−0.061	−11.6	−0.172	0.013	−0.063	−13.3
−3.0	−0.127	0.019	−0.062	−6.6	−0.131	0.011	−0.063	−12.2
−2.5	−0.069	0.018	−0.064	−3.8	−0.075	0.011	−0.063	−6.8
−2.0	−0.014	0.018	−0.064	−0.8	−0.020	0.011	−0.064	−1.9
−1.5	0.041	0.018	−0.064	2.3	0.007	0.011	−0.064	0.7
−1.0	0.095	0.017	−0.064	5.5	0.090	0.011	−0.064	8.3
−0.5	0.148	0.017	−0.064	8.8	0.146	0.011	−0.064	13.4
0.0	0.203	0.017	−0.064	12.2	0.201	0.011	−0.065	18.6
0.5	0.256	0.017	−0.064	15.5	0.257	0.011	−0.065	23.8
1.0	0.311	0.016	−0.064	18.9	0.312	0.011	−0.065	28.7
1.5	0.365	0.016	−0.065	22.3	0.368	0.011	−0.065	33.8
2.0	0.419	0.016	−0.065	25.4	0.423	0.011	−0.066	38.5
2.5	0.474	0.017	−0.065	28.7	0.479	0.011	−0.066	43.1
3.0	0.528	0.017	−0.065	31.7	0.534	0.011	−0.066	47.7
3.5	0.583	0.017	−0.065	34.7	0.589	0.011	−0.066	51.8
4.0	0.638	0.017	−0.065	37.4	0.644	0.012	−0.066	55.8
4.5	0.692	0.017	−0.065	40.0	0.698	0.012	−0.066	59.5
5.0	0.746	0.018	−0.065	42.4	0.752	0.012	−0.066	62.9
5.5	0.798	0.018	−0.065	44.7	0.805	0.012	−0.066	66.2
6.0	0.850	0.018	−0.064	46.8	0.857	0.012	−0.066	69.0
6.5	0.901	0.018	−0.064	48.8	0.909	0.013	−0.065	71.5
7.0	0.950	0.019	−0.063	50.6	0.960	0.013	−0.065	73.8
7.5	0.996	0.019	−0.061	52.3	1.010	0.013	−0.064	75.8
8.0	1.040	0.019	−0.060	53.7	1.059	0.014	−0.063	77.2
8.5	1.082	0.020	−0.057	54.9	1.103	0.014	−0.062	77.6
9.0	1.117	0.020	−0.054	55.6	1.140	0.015	−0.059	76.0
9.5	1.145	0.021	−0.050	55.5	1.171	0.017	−0.051	68.0
10.0	1.160	0.022	−0.044	53.9	1.191	0.019	−0.046	63.0
10.5	1.165	0.023	−0.037	49.8	1.200	0.020	−0.044	60.7

(*Continued*)

TABLE 12.10 (*Continued*)

Aerodynamic Coefficients of NREL S819 Airfoil

	Re = 200,000				Re = 500,000			
α	C_z	C_x	C_m	f	C_z	C_x	C_m	f
11.0	1.161	0.026	−0.031	44.2	1.219	0.022	−0.040	56.6
11.3	1.159	0.028	−0.029	41.5	1.227	0.023	−0.038	54.5
11.5	1.157	0.030	−0.026	38.9	1.233	0.024	−0.036	52.0
12.0	1.153	0.034	−0.023	34.2	1.245	0.026	−0.033	47.3
12.5	1.148	0.038	−0.021	29.9	1.254	0.029	−0.030	42.7
13.0	1.144	0.043	−0.019	26.3	1.262	0.033	−0.028	38.2
13.5	1.135	0.048	−0.017	23.5	1.266	0.037	−0.027	34.0
14.0	1.128	0.054	−0.017	20.9	1.265	0.043	−0.026	29.7
14.5	1.126	0.061	−0.019	18.6	1.259	0.049	−0.026	25.9
15.0	1.126	0.067	−0.021	16.7	1.249	0.054	−0.026	23.0
15.5	1.126	0.074	−0.023	15.2	1.237	0.062	−0.028	19.8
16.0	1.128	0.081	−0.025	13.9	1.236	0.070	−0.030	17.8
16.5	1.127	0.089	−0.028	12.7	1.221	0.079	−0.034	15.5
17.0	1.130	0.096	−0.030	11.8	1.219	0.087	−0.037	14.0
17.5	1.124	0.105	−0.035	10.8	1.207	0.096	−0.041	12.5
18.0	1.128	0.111	−0.038	10.1	1.200	0.102	−0.043	11.8
18.5	1.116	0.122	−0.043	9.1				

Blank cells indicate that the airfoil is stalled.
N.B.: the pitching moment is referred to x = 0.25; y = 0.

12.4.6 GOE 417-A Airfoil (Cambered Plate)

This is a concave–convex airfoil, featuring rather mediocre aerodynamic performance, but it is very suitable for the quick construction of multi-blade rotors, employing curve blades cut out of plastic tube. The relative thickness is 3.2% (Figure 12.8; Tables 12.11 and 12.12).

12.4.7 Airfoil Eppler E377 (Modified)

This is a version with reduced relative thickness (3.8%) of the E377 airfoil, very suitable for the construction of profiled semi-rigid sails or of profiles in

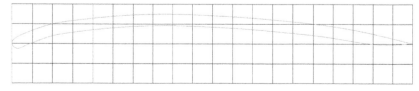

FIGURE 12.8
Geometry of the GOE 417-A airfoil.

TABLE 12.11

Coordinates of the GOE 417A Airfoil

Extrados		Intrados	
x	y	x	y
0	0	0	0
0.0125	0.0155	0.0125	−0.014
0.025	0.022	0.025	−0.01
0.05	0.0325	0.05	0.001
0.075	0.0415	0.075	0.0105
0.1	0.0485	0.1	0.0185
0.15	0.057	0.15	0.0275
0.2	0.063	0.2	0.034
0.3	0.0715	0.3	0.0425
0.4	0.0735	0.4	0.0445
0.5	0.07	0.5	0.041
0.6	0.064	0.6	0.035
0.7	0.0545	0.7	0.0255
0.8	0.0425	0.8	0.0135
0.9	0.024	0.9	−0.002
1	0	1	0

TABLE 12.12

Aerodynamic Coefficients of the GOE 417A Airfoil

α	Re = 200,000				Re = 500,000			
	C_z	C_x	C_m	f	C_z	C_x	C_m	f
−9.5					−0.371	0.114	0.112	−3.2
−9					−0.372	0.113	0.110	−3.3
−8.5					−0.371	0.106	0.104	−3.5
−8					−0.356	0.102	0.100	−3.5
−7.5	−0.293	0.092	−0.013	−3.2	−0.340	0.098	0.096	−3.4
−7	−0.303	0.089	−0.009	−3.4	−0.303	0.092	0.090	−3.3
−6.5	−0.314	0.085	−0.006	−3.7	−0.253	0.085	0.083	−3.0
−6	−0.316	0.081	−0.005	−3.9	−0.209	0.079	0.077	−2.6
−5.5	−0.280	0.075	−0.013	−3.7	−0.133	0.073	0.070	−1.8
−5	−0.320	0.082	−0.011	−3.9	−0.044	0.079	0.076	−0.6
−4.5	−0.227	0.076	−0.031	−3.0	0.011	0.062	0.059	0.2
−4	−0.178	0.067	−0.037	−2.6	0.066	0.046	0.043	1.4
−3.5	−0.112	0.061	−0.046	−1.8	0.221	0.043	0.040	5.1
−3	−0.038	0.056	−0.056	−0.7	0.243	0.031	0.028	7.8
−2.5	0.063	0.052	−0.070	1.2	0.266	0.019	0.016	14.0
−2	0.117	0.045	−0.074	2.6	0.327	0.015	0.012	22.6

(Continued)

TABLE 12.12 (*Continued*)

Aerodynamic Coefficients of the GOE 417A Airfoil

	Re = 200,000				Re = 500,000			
α	C_z	C_x	C_m	f	C_z	C_x	C_m	f
−1.5	0.184	0.041	−0.079	4.5	0.378	0.012	0.009	32.1
−1	0.279	0.039	−0.086	7.1	0.442	0.010	0.007	43.3
−0.5	0.338	0.033	−0.091	10.3	0.493	0.008	0.004	62.4
0	0.418	0.030	−0.094	13.7	0.550	0.020	0.017	27.1
0.5	0.485	0.026	−0.097	18.5	0.599	0.018	0.014	33.0
1	0.564	0.025	−0.099	22.7	0.653	0.016	0.012	39.8
1.5	0.625	0.020	−0.101	30.7	0.701	0.015	0.010	47.0
2	0.693	0.020	−0.101	35.3	0.743	0.014	0.009	53.0
2.5	0.761	0.016	−0.103	46.8	0.786	0.013	0.008	59.0
3	0.826	0.015	−0.105	55.7	0.824	0.013	0.007	64.0
3.5	0.879	0.014	−0.104	63.3	0.864	0.013	0.006	66.4
4	0.928	0.014	−0.101	67.9	0.904	0.013	0.006	68.8
4.5	0.974	0.014	−0.097	69.4	0.944	0.013	0.006	71.1
5	1.014	0.014	−0.093	71.3	0.983	0.013	0.006	73.3
5.5	1.047	0.015	−0.088	71.2	1.023	0.014	0.005	75.6
6	1.077	0.016	−0.083	65.6	1.058	0.014	0.006	73.4
6.5	1.114	0.018	−0.080	62.9	1.089	0.015	0.007	70.3
7	1.226	0.021	−0.095	58.3	1.123	0.016	0.008	68.2
7.5	1.244	0.024	−0.088	52.3	1.296	0.019	0.011	68.6
8	1.274	0.026	−0.083	49.1	1.326	0.020	0.013	65.4
8.5	1.308	0.028	−0.079	46.0	1.347	0.023	0.015	59.5
9	1.359	0.033	−0.079	40.7	1.374	0.024	0.017	56.3
9.5	1.401	0.039	−0.077	36.2	1.399	0.026	0.019	53.5
10	1.423	0.042	−0.071	34.1	1.422	0.028	0.021	50.5
10.5	1.440	0.048	−0.065	30.2	1.446	0.031	0.024	47.1
11	1.384	0.046	−0.050	30.3	1.476	0.036	0.029	41.4
11.5	1.330	0.064	−0.053	20.8	1.482	0.038	0.032	38.5
12	1.277	0.082	−0.057	15.5	1.484	0.042	0.037	35.2
12.5	1.223	0.101	−0.060	12.2	1.473	0.047	0.042	31.1
13	1.169	0.119	−0.063	9.8	1.449	0.054	0.049	27.0
13.5	1.116	0.137	−0.067	8.1	1.415	0.061	0.057	23.2
14	1.062	0.155	−0.070	6.8	1.374	0.070	0.067	19.5
14.5	1.076	0.160	−0.069	6.7	1.328	0.082	0.078	16.2
15					1.282	0.095	0.092	13.4
15.5					1.234	0.113	0.110	11.0
16					1.188	0.132	0.129	9.0

N.B.: the pitching moment is referred to $x = 0.25$; $y = 0$.

curved metallic sheet, with their leading edge made out of plastic or wood. Like any other thin profile, it offers good aerodynamic performance with low Re (Figure 12.9; Tables 12.13 and 12.14).

12.4.8 Simmetric Airfoil Eppler E169 (14.4%)

This is a symmetric airfoil, suitable for the construction of vertical axis wind turbines. It is specially conceived to offer good aerodynamic performance with low Re (Figure 12.10; Tables 12.15 and 12.16).

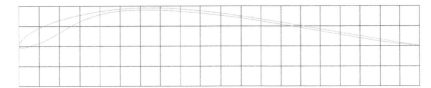

FIGURE 12.9
Geometry of the E377 MOD airfoil.

TABLE 12.13

Coordinates of the E377 MOD Airfoil

Extrados		Intrados	
x	y	x	y
0.0001	−0.00102	0	0
0.00136	0.00482	0.0001	−0.00102
0.00602	0.01244	0.00324	−0.00406
0.01428	0.02102	0.01141	−0.00463
0.02625	0.03013	0.02442	−0.00204
0.04196	0.03945	0.0429	0.00363
0.06134	0.04871	0.065	0.0125
0.08428	0.05767	0.08758	0.02431
0.11062	0.06611	0.11227	0.03916
0.14014	0.07381	0.14	0.055
0.17259	0.08059	0.17176	0.06882
0.20766	0.08625	0.2077	0.0793
0.24503	0.09054	0.245	0.0855
0.28446	0.09316	0.2845	0.0882
0.32584	0.09397	0.32634	0.0886
0.36901	0.09292	0.36956	0.08713
0.41379	0.09006	0.4138	0.0841
0.45998	0.08553	0.461	0.07964
0.50741	0.07959	0.50844	0.07411
0.55567	0.07272	0.5557	0.0677
0.60416	0.06527	0.60486	0.06027

(Continued)

TABLE 12.13 (*Continued*)

Coordinates of the E377 MOD Airfoil

Extrados		Intrados	
x	y	x	y
0.65229	0.05751	0.65291	0.05249
0.69944	0.04966	0.6994	0.0447
0.74501	0.04193	0.74539	0.03692
0.78841	0.03449	0.78878	0.0295
0.82909	0.02751	0.8291	0.0225
0.8665	0.02112	0.86668	0.01597
0.90015	0.01545	0.90029	0.01058
0.92957	0.0106	0.9296	0.0066
0.95434	0.00664	0.95452	0.00391
0.97407	0.00363	0.97423	0.00212
0.98841	0.00159	0.98849	0.00093
0.9971	0.0004	0.99711	0.00023
1	0	1	0

TABLE 12.14

Aerodynamic Coefficients of the E377 MOD Airfoil

α	C_z	C_x	C_m	f
Re = 200,000				
−6.50	−0.08500	0.08678	−0.04110	−0.98
−6.25	−0.05670	0.08527	−0.04690	−0.66
−6.00	−0.03680	0.08388	−0.05030	−0.44
−5.75	−0.01610	0.08239	−0.05370	−0.20
−5.50	−0.00580	0.07764	−0.05350	−0.07
−5.25	0.00200	0.07321	−0.05230	0.03
−5.00	0.01820	0.07053	−0.05360	0.26
−4.75	0.03760	0.06807	−0.05570	0.55
−4.50	0.05950	0.06574	−0.05830	0.91
−4.25	0.08570	0.06351	−0.06170	1.35
−4.00	0.12760	0.06300	−0.06840	2.03
−3.75	0.16830	0.06222	−0.07480	2.70
−3.50	0.19710	0.05917	−0.07840	3.33
−3.25	0.20460	0.05395	−0.07710	3.79
−3.00	0.22690	0.05121	−0.07860	4.43
−2.75	0.25530	0.04896	−0.08140	5.21
−2.50	0.28670	0.04685	−0.08440	6.12
−2.25	0.33120	0.04592	−0.08950	7.21
−2.00	0.37570	0.04430	−0.09480	8.48
−1.75	0.39330	0.04061	−0.09520	9.68

(Continued)

TABLE 12.14 (*Continued*)

Aerodynamic Coefficients of the E377 MOD Airfoil

α	C_z	C_x	C_m	f
−1.50	0.42190	0.03864	−0.09700	10.92
−1.25	0.45490	0.03694	−0.09950	12.31
−1.00	0.50180	0.03702	−0.10320	13.55
−0.75	0.53990	0.03535	−0.10610	15.27
−0.50	0.55950	0.03213	−0.10660	17.41
−0.25	0.58960	0.03087	−0.10780	19.10
0.00	0.63810	0.03213	−0.11030	19.86
0.25	0.65980	0.02852	−0.11120	23.13
0.50	0.68770	0.02710	−0.11210	25.38
0.75	0.71910	0.02617	−0.11300	27.48
1.00	0.75980	0.02675	−0.11380	28.40
1.25	0.78310	0.02418	−0.11460	32.39
1.50	0.81840	0.02477	−0.11470	33.04
1.75	0.84900	0.02343	−0.11530	36.24
2.00	0.87600	0.02182	−0.11580	40.15
2.25	0.90450	0.02105	−0.11590	42.97
2.50	0.93660	0.02134	−0.11560	43.89
2.75	0.96440	0.02027	−0.11590	47.58
3.00	0.99260	0.01989	−0.11570	49.90
3.25	1.02010	0.01958	−0.11570	52.10
3.50	1.04695	0.01914	−0.11565	54.71
3.75	1.07380	0.01869	−0.11560	57.45
4.00	1.10080	0.01871	−0.11540	58.83
4.25	1.12730	0.01835	−0.11530	61.43
4.50	1.15400	0.01821	−0.11510	63.37
4.75	1.17960	0.01802	−0.11500	65.46
5.00	1.20560	0.01755	−0.11480	68.70
5.25	1.23180	0.01739	−0.11460	70.83
5.50	1.25820	0.01736	−0.11420	72.48
5.75	1.28450	0.01736	−0.11380	73.99
6.00	1.31070	0.01756	−0.11320	74.64
6.25	1.33730	0.01777	−0.11270	75.26
6.50	1.36290	0.01789	−0.11210	76.18
6.75	1.38870	0.01795	−0.11150	77.36
7.00	1.41390	0.01798	−0.11090	78.64
7.25	1.43930	0.01801	−0.11040	79.92
7.50	1.46460	0.01799	−0.10990	81.41
7.75	1.48860	0.01801	−0.10920	82.65
8.00	1.51250	0.01803	−0.10850	83.89
8.25	1.53610	0.01808	−0.10780	84.96

(*Continued*)

TABLE 12.14 (*Continued*)

Aerodynamic Coefficients of the E377 MOD Airfoil

α	C_z	C_x	C_m	f
8.50	1.55930	0.01818	−0.10700	85.77
8.75	1.58180	0.01832	−0.10620	86.34
9.00	1.60290	0.01855	−0.10520	86.41
9.25	1.62130	0.01903	−0.10400	85.20
9.50	1.63650	0.01999	−0.10260	81.87
9.75	1.64620	0.02153	−0.10070	76.46
10.00	1.65330	0.02329	−0.09870	70.99
10.25	1.65810	0.02512	−0.09650	66.01
10.50	1.66100	0.02693	−0.09400	61.68
10.75	1.65900	0.02876	−0.09080	57.68
11.00	1.64880	0.03118	−0.08730	52.88
11.25	1.63240	0.03462	−0.08460	47.15
11.50	1.61770	0.03850	−0.08290	42.02
11.75	1.60560	0.04256	−0.08200	37.73
12.00	1.59950	0.04617	−0.08150	34.64
12.25	1.59230	0.05007	−0.08130	31.80
12.50	1.58390	0.05406	−0.08155	29.30
12.75	1.57550	0.05805	−0.08180	27.14
13.00	1.56380	0.06286	−0.08210	24.88
13.25	1.55220	0.06770	−0.08240	22.93
13.50	1.54000	0.07268	−0.08270	21.19
13.75	1.53580	0.07667	−0.08280	20.03
14.00	1.53390	0.08045	−0.08310	19.07
14.25	1.53210	0.08432	−0.08350	18.17
14.50	1.52990	0.08832	−0.08410	17.32
14.75	1.52750	0.09239	−0.08470	16.53
15.00	1.52540	0.09641	−0.08510	15.82
15.25	1.52370	0.10046	−0.08560	15.17
Re = 500,000				
−6.50	−0.09570	0.08424	−0.03700	−1.14
−6.25	−0.08370	0.08132	−0.03790	−1.03
−6.00	−0.06880	0.07889	−0.03970	−0.87
−5.50	−0.03430	0.07410	−0.04410	−0.46
−5.50	−0.02460	0.07288	−0.04540	−0.34
−5.25	−0.01490	0.07166	−0.04670	−0.21
−5.00	0.00590	0.06919	−0.04940	0.09
−4.75	0.02810	0.06665	−0.05240	0.42
−4.50	0.05869	0.06399	−0.05589	0.92
−4.25	0.08930	0.06150	−0.05940	1.45
−4.00	0.11991	0.05900	−0.06291	2.03

(*Continued*)

TABLE 12.14 (*Continued*)

Aerodynamic Coefficients of the E377 MOD Airfoil

α	C_z	C_x	C_m	f
−3.75	0.15052	0.05651	−0.06642	2.66
−3.50	0.18113	0.05401	−0.06993	3.35
−3.25	0.21173	0.05152	−0.07343	4.11
−3.00	0.24234	0.04903	−0.07694	4.94
−2.75	0.27295	0.04653	−0.08045	5.87
−2.50	0.30356	0.04404	−0.08396	6.89
−2.25	0.33417	0.04154	−0.08747	8.04
−2.00	0.36478	0.03905	−0.09098	9.34
−1.75	0.39540	0.03672	−0.09450	10.77
−1.50	0.42450	0.03511	−0.09630	12.09
−1.25	0.45850	0.03343	−0.09870	13.72
−1.00	0.50100	0.03265	−0.10180	15.34
−0.75	0.53600	0.03113	−0.10390	17.22
−0.50	0.56960	0.02786	−0.10660	20.45
−0.25	0.59570	0.02648	−0.10740	22.50
0.00	0.62640	0.02532	−0.10850	24.74
0.25	0.66560	0.02513	−0.10980	26.49
0.50	0.69590	0.02325	−0.11100	29.94
0.75	0.72620	0.02136	−0.11220	34.00
1.00	0.75540	0.02040	−0.11270	37.03
1.25	0.78590	0.01945	−0.11320	40.41
1.50	0.82520	0.01657	−0.11440	49.80
1.75	0.86370	0.01157	−0.11560	74.65
2.00	0.89150	0.01101	−0.11540	80.97
2.25	0.91860	0.01146	−0.11500	80.16
2.50	0.94490	0.01251	−0.11460	75.53
2.75	0.97150	0.01283	−0.11440	75.72
3.00	0.99740	0.01329	−0.11410	75.05
3.25	1.02440	0.01325	−0.11400	77.31
3.50	1.05150	0.01309	−0.11380	80.33
3.75	1.07900	0.01281	−0.11350	84.23
4.00	1.10520	0.01288	−0.11340	85.81
4.25	1.13220	0.01348	−0.11290	83.99
4.50	1.15840	0.01286	−0.11300	90.08
4.75	1.18510	0.01269	−0.11290	93.39
5.00	1.21190	0.01274	−0.11250	95.13
5.25	1.23820	0.01254	−0.11240	98.74
5.50	1.26480	0.01255	−0.11210	100.78
5.75	1.29090	0.01245	−0.11190	103.69
6.00	1.31730	0.01256	−0.11160	104.88

(*Continued*)

TABLE 12.14 (*Continued*)

Aerodynamic Coefficients of the E377 MOD Airfoil

α	C_z	C_x	C_m	f
6.25	1.34310	0.01244	−0.11140	107.97
6.50	1.36920	0.01252	−0.11110	109.36
6.75	1.39490	0.01246	−0.11090	111.95
7.00	1.42040	0.01255	−0.11060	113.18
7.25	1.44590	0.01264	−0.11030	114.39
7.50	1.47190	0.01256	−0.10970	117.19
7.75	1.49680	0.01272	−0.10930	117.67
8.00	1.52050	0.01296	−0.10890	117.32
8.25	1.54280	0.01345	−0.10830	114.71
8.50	1.56230	0.01435	−0.10750	108.87
8.75	1.57900	0.01560	−0.10640	101.22
9.00	1.59640	0.01672	−0.10540	95.48
9.25	1.61320	0.01781	−0.10430	90.58
9.50	1.62770	0.01906	−0.10300	85.40
9.75	1.63880	0.02058	−0.10120	79.63
10.00	1.64230	0.02266	−0.09860	72.48
10.25	1.65060	0.02406	−0.09630	68.60
10.50	1.65890	0.02527	−0.09400	65.65
10.75	1.66070	0.02650	−0.09070	62.67
11.00	1.65880	0.02813	−0.08750	58.97
11.25	1.65750	0.03002	−0.08510	55.21
11.50	1.65680	0.03213	−0.08340	51.57
11.75	1.65530	0.03459	−0.08210	47.85
12.00	1.65300	0.03737	−0.08110	44.23
12.25	1.64910	0.04055	−0.08050	40.67
12.50	1.64460	0.04396	−0.08020	37.41
12.75	1.63920	0.04761	−0.08010	34.43
13.00	1.63250	0.05156	−0.08010	31.66
13.25	1.62550	0.05564	−0.08030	29.21
13.50	1.61760	0.05997	−0.08070	26.97
13.75	1.61000	0.06432	−0.08120	25.03
14.00	1.60200	0.06890	−0.08190	23.25
14.25	1.59380	0.07367	−0.08280	21.63
14.50	1.58670	0.07843	−0.08370	20.23
14.75	1.58010	0.08319	−0.08480	18.99
15.00	1.57410	0.08799	−0.08590	17.89
15.25	1.56840	0.09283	−0.08700	16.90

N.B.: the pitching moment is referred to $x = 0.25$; $y = 0$.

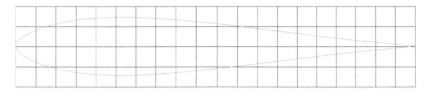

FIGURE 12.10
Geometry of the E169 (14.4%) airfoil.

TABLE 12.15

Coordinates of the E169 (14.4%) Airfoil

x	y Extrados	y Intrados
0.000000	0.000000	0.000000
0.002900	0.008190	−0.008190
0.011060	0.017290	−0.017290
0.023790	0.026570	−0.026570
0.040850	0.035670	−0.035670
0.062060	0.044270	−0.044270
0.087210	0.052120	−0.052120
0.116050	0.058960	−0.058960
0.148390	0.064500	−0.064500
0.184080	0.068580	−0.068580
0.222890	0.071100	−0.071100
0.264560	0.072020	−0.072020
0.308840	0.071270	−0.071270
0.355550	0.068890	−0.068890
0.404420	0.065160	−0.065160
0.454990	0.060360	−0.060360
0.506750	0.054770	−0.054770
0.559120	0.048660	−0.048660
0.611480	0.042310	−0.042310
0.663160	0.035960	−0.035960
0.713460	0.029850	−0.029850
0.761680	0.024160	−0.024160
0.807080	0.019040	−0.019040
0.848990	0.014580	−0.014580
0.886730	0.010810	−0.010810
0.919700	0.007710	−0.007710
0.947370	0.005140	−0.005140
0.969480	0.002900	−0.002900
0.985980	0.001150	−0.001150
0.996400	0.000220	−0.000220
1.000000	0.000000	0.000000

TABLE 12.16

Aerodynamic Coefficients of the E169 (14.4%) Airfoil

α	C_z	C_x	C_m	f
Re = 100,000				
−15.00	−0.9141	0.09832	−0.01	−9.30
−14.75	−0.95	0.09	−0.02	−10.65
−14.50	−0.98	0.08	−0.02	−12.06
−14.25	−1.00	0.07	−0.03	−13.51
−14.00	−1.03	0.07	−0.03	−14.98
−13.75	−1.05	0.06	−0.03	−16.49
−13.50	−1.06	0.06	−0.03	−18.01
−13.25	−1.08	0.06	−0.03	−19.48
−13.00	−1.08	0.05	−0.03	−20.77
−12.75	−1.07	0.05	−0.03	−21.59
−12.50	−1.07	0.05	−0.03	−22.81
−12.25	−1.07	0.04	−0.03	−24.14
−12.00	−1.07	0.04	−0.03	−25.54
−11.75	−1.06	0.04	−0.03	−26.42
−11.50	−1.06	0.04	−0.03	−27.71
−11.25	−1.04	0.04	−0.03	−28.94
−11.00	−1.04	0.03	−0.02	−30.23
−10.75	−1.03	0.03	−0.02	−31.56
−10.50	−1.02	0.03	−0.02	−33.04
−10.25	−1.02	0.03	−0.02	−34.50
−10.00	−1.01	0.03	−0.02	−35.87
−9.75	−1.00	0.03	−0.01	−37.14
−9.50	−0.99	0.03	−0.01	−38.17
−9.25	−0.97	0.03	−0.01	−38.80
−9.00	−0.96	0.02	−0.01	−39.30
−8.75	−0.94	0.02	−0.01	−39.22
−8.50	−0.92	0.02	−0.01	−39.00
−8.25	−0.90	0.02	−0.01	−38.73
−8.00	−0.88	0.02	0.00	−38.36
−7.75	−0.86	0.02	0.00	−37.84
−7.50	−0.84	0.02	0.00	−37.16
−7.25	−0.82	0.02	0.00	−36.41
−7.00	−0.79	0.02	0.00	−35.67
−6.75	−0.77	0.02	0.00	−35.00
−6.50	−0.75	0.02	0.00	−34.45
−6.25	−0.73	0.02	0.00	−34.01
−6.00	−0.70	0.02	0.01	−33.44
−5.75	−0.68	0.02	0.01	−32.64
−5.50	−0.66	0.02	0.01	−32.13

(Continued)

TABLE 12.16 (*Continued*)

Aerodynamic Coefficients of the E169 (14.4%) Airfoil

α	C_z	C_x	C_m	f
−5.25	−0.64	0.02	0.01	−31.77
−5.00	−0.62	0.02	0.01	−31.07
−4.75	−0.60	0.02	0.01	−30.62
−4.50	−0.59	0.02	0.01	−30.39
−4.25	−0.57	0.02	0.01	−29.91
−4.00	−0.55	0.02	0.01	−28.70
−3.75	−0.49	0.02	0.01	−25.96
−3.50	−0.43	0.02	0.00	−23.33
−3.25	−0.38	0.02	0.00	−20.73
−3.00	−0.33	0.02	−0.01	−18.41
−2.75	−0.29	0.02	−0.01	−16.25
−2.50	−0.26	0.02	−0.01	−14.47
−2.25	−0.23	0.02	−0.01	−12.92
−2.00	−0.20	0.02	−0.01	−11.43
−1.75	−0.18	0.02	−0.01	−10.01
−1.50	−0.15	0.02	−0.01	−8.61
−1.25	−0.13	0.02	0.00	−7.20
−1.00	−0.10	0.02	0.00	−5.76
−0.75	−0.08	0.02	0.00	−4.33
−0.50	−0.05	0.02	0.00	−2.90
−0.25	−0.03	0.02	0.00	−1.45
0.00	0.00	0.02	0.00	0.00
0.25	0.03	0.02	0.00	1.45
0.50	0.05	0.02	0.00	2.90
0.75	0.08	0.02	0.00	4.33
1.00	0.10	0.02	0.00	5.76
1.25	0.13	0.02	0.00	7.20
1.50	0.15	0.02	0.01	8.61
1.75	0.18	0.02	0.01	10.01
2.00	0.20	0.02	0.01	11.43
2.25	0.23	0.02	0.01	12.92
2.50	0.26	0.02	0.01	14.47
2.75	0.29	0.02	0.01	16.25
3.00	0.33	0.02	0.01	18.41
3.25	0.38	0.02	0.00	20.73
3.50	0.43	0.02	0.00	23.33
3.75	0.49	0.02	−0.01	25.97
4.00	0.55	0.02	−0.01	28.70
4.25	0.57	0.02	−0.01	29.91
4.50	0.59	0.02	−0.01	30.41
4.75	0.60	0.02	−0.01	30.62
5.00	0.62	0.02	−0.01	31.09

(*Continued*)

TABLE 12.16 (*Continued*)

Aerodynamic Coefficients of the E169 (14.4%) Airfoil

α	C_z	C_x	C_m	f
5.25	0.64	0.02	−0.01	31.78
5.50	0.66	0.02	−0.01	32.16
5.75	0.68	0.02	−0.01	32.65
6.00	0.70	0.02	−0.01	33.44
6.25	0.73	0.02	0.00	34.01
6.50	0.75	0.02	0.00	34.46
6.75	0.77	0.02	0.00	35.00
7.00	0.79	0.02	0.00	35.67
7.25	0.82	0.02	0.00	36.41
7.50	0.84	0.02	0.00	37.16
7.75	0.86	0.02	0.00	37.84
8.00	0.88	0.02	0.00	38.37
8.25	0.90	0.02	0.01	38.73
8.50	0.92	0.02	0.01	39.01
8.75	0.94	0.02	0.01	39.22
9.00	0.96	0.02	0.01	39.29
9.25	0.97	0.03	0.01	38.80
9.50	0.99	0.03	0.01	38.19
9.75	1.00	0.03	0.01	37.14
10.00	1.01	0.03	0.02	35.88
10.25	1.02	0.03	0.02	34.50
10.50	1.02	0.03	0.02	33.05
10.75	1.03	0.03	0.02	31.57
11.00	1.04	0.03	0.02	30.24
11.25	1.04	0.04	0.03	28.95
11.50	1.06	0.04	0.03	27.72
11.75	1.06	0.04	0.03	26.42
12.00	1.07	0.04	0.03	25.55
12.25	1.07	0.04	0.03	24.16
12.50	1.07	0.05	0.03	22.83
12.75	1.07	0.05	0.03	21.61
13.00	1.08	0.05	0.03	20.81
13.25	1.08	0.06	0.03	19.52
13.50	1.06	0.06	0.03	18.03
13.75	1.05	0.06	0.03	16.52
14.00	1.03	0.07	0.03	15.01
14.25	1.00	0.07	0.03	13.50
14.50	0.98	0.08	0.02	12.06
14.75	0.95	0.09	0.02	10.58
15.00	0.91	0.10	0.01	9.28
15.25	0.70	0.16	−0.02	4.43
15.50	0.72	0.16	−0.02	4.43

(*Continued*)

TABLE 12.16 (*Continued*)

Aerodynamic Coefficients of the E169 (14.4%) Airfoil

α	C_z	C_x	C_m	f
Re = 200,000				
−17.75	−0.92	0.14	0.01	−6.75
−17.50	−0.94	0.13	0.01	−7.50
−17.25	−0.97	0.12	0.00	−8.42
−17.00	−0.99	0.11	−0.01	−9.26
−16.75	−1.01	0.10	−0.01	−10.17
−16.50	−1.03	0.09	−0.01	−11.10
−16.25	−1.05	0.09	−0.02	−12.08
−16.00	−1.06	0.08	−0.02	−13.13
−15.75	−1.07	0.08	−0.02	−13.95
−15.50	−1.07	0.07	−0.02	−14.54
−15.25	−1.07	0.07	−0.02	−15.36
−15.00	−1.07	0.07	−0.03	−16.23
−14.75	−1.07	0.06	−0.03	−17.21
−14.50	−1.08	0.06	−0.03	−18.22
−14.25	−1.08	0.06	−0.03	−19.27
−14.00	−1.08	0.05	−0.03	−20.35
−13.75	−1.07	0.05	−0.03	−21.35
−13.50	−1.07	0.05	−0.03	−22.36
−13.25	−1.07	0.05	−0.03	−23.50
−13.00	−1.07	0.04	−0.03	−24.82
−12.75	−1.06	0.04	−0.03	−26.14
−12.50	−1.06	0.04	−0.03	−27.64
−12.25	−1.06	0.04	−0.03	−29.37
−12.00	−1.06	0.03	−0.03	−31.06
−11.75	−1.05	0.03	−0.03	−32.68
−11.50	−1.05	0.03	−0.03	−35.15
−11.25	−1.05	0.03	−0.03	−37.24
−11.00	−1.05	0.03	−0.02	−38.86
−10.75	−1.05	0.03	−0.02	−41.50
−10.50	−1.04	0.02	−0.02	−43.11
−10.25	−1.04	0.02	−0.02	−45.57
−10.00	−1.03	0.02	−0.01	−47.84
−9.75	−1.02	0.02	−0.01	−49.82
−9.50	−1.00	0.02	−0.01	−51.90
−9.25	−0.99	0.02	−0.01	−53.44
−9.00	−0.97	0.02	−0.01	−54.63
−8.75	−0.95	0.02	−0.01	−55.47
−8.50	−0.93	0.02	−0.01	−55.83
−8.25	−0.91	0.02	0.00	−56.07

(*Continued*)

TABLE 12.16 (*Continued*)

Aerodynamic Coefficients of the E169 (14.4%) Airfoil

α	C_z	C_x	C_m	f
−8.00	−0.89	0.02	0.00	−56.08
−7.75	−0.87	0.02	0.00	−55.84
−7.50	−0.85	0.02	0.00	−55.49
−7.25	−0.82	0.01	0.00	−54.95
−7.00	−0.80	0.01	0.00	−54.31
−6.75	−0.78	0.01	0.00	−53.53
−6.50	−0.76	0.01	0.00	−52.81
−6.25	−0.73	0.01	0.00	−52.03
−6.00	−0.71	0.01	0.00	−51.15
−5.75	−0.69	0.01	0.01	−50.06
−5.50	−0.67	0.01	0.01	−49.21
−5.25	−0.65	0.01	0.01	−48.30
−5.00	−0.62	0.01	0.01	−47.09
−4.75	−0.58	0.01	0.00	−43.90
−4.50	−0.53	0.01	0.00	−40.90
−4.25	−0.48	0.01	0.00	−38.09
−4.00	−0.44	0.01	−0.01	−34.98
−3.75	−0.40	0.01	−0.01	−32.21
−3.50	−0.37	0.01	−0.01	−29.93
−3.25	−0.34	0.01	−0.01	−27.78
−3.00	−0.32	0.01	−0.01	−25.82
−2.75	−0.29	0.01	−0.01	−23.76
−2.50	−0.27	0.01	0.00	−21.67
−2.25	−0.24	0.01	0.00	−19.59
−2.00	−0.21	0.01	0.00	−17.45
−1.75	−0.19	0.01	0.00	−15.35
−1.50	−0.16	0.01	0.00	−13.19
−1.25	−0.13	0.01	0.00	−11.01
−1.00	−0.11	0.01	0.00	−8.84
−0.75	−0.08	0.01	0.00	−6.63
−0.50	−0.05	0.01	0.00	−4.43
−0.25	−0.03	0.01	0.00	−2.22
0.00	0.00	0.01	0.00	0.00
0.25	0.03	0.01	0.00	2.22
0.50	0.05	0.01	0.00	4.43
0.75	0.08	0.01	0.00	6.62
1.00	0.11	0.01	0.00	8.84
1.25	0.13	0.01	0.00	11.01
1.50	0.16	0.01	0.00	13.19
1.75	0.19	0.01	0.00	15.35

(*Continued*)

TABLE 12.16 (*Continued*)

Aerodynamic Coefficients of the E169 (14.4%) Airfoil

α	C_z	C_x	C_m	f
2.00	0.21	0.01	0.00	17.45
2.25	0.24	0.01	0.00	19.59
2.50	0.27	0.01	0.00	21.67
2.75	0.29	0.01	0.01	23.75
3.00	0.32	0.01	0.01	25.79
3.25	0.34	0.01	0.01	27.78
3.50	0.37	0.01	0.01	29.95
3.75	0.40	0.01	0.01	32.21
4.00	0.44	0.01	0.01	34.98
4.25	0.48	0.01	0.00	38.09
4.50	0.53	0.01	0.00	40.90
4.75	0.58	0.01	0.00	43.89
5.00	0.62	0.01	−0.01	47.09
5.25	0.65	0.01	−0.01	48.31
5.50	0.67	0.01	−0.01	49.22
5.75	0.69	0.01	−0.01	50.07
6.00	0.71	0.01	0.00	51.16
6.25	0.73	0.01	0.00	52.04
6.50	0.76	0.01	0.00	52.82
6.75	0.78	0.01	0.00	53.54
7.00	0.80	0.01	0.00	54.35
7.25	0.82	0.01	0.00	54.95
7.50	0.85	0.02	0.00	55.49
7.75	0.87	0.02	0.00	55.84
8.00	0.89	0.02	0.00	56.08
8.25	0.91	0.02	0.00	56.07
8.50	0.93	0.02	0.01	55.83
8.75	0.95	0.02	0.01	55.47
9.00	0.97	0.02	0.01	54.63
9.25	0.99	0.02	0.01	53.44
9.50	1.00	0.02	0.01	51.89
9.75	1.02	0.02	0.01	49.81
10.00	1.03	0.02	0.01	47.84
10.25	1.04	0.02	0.02	45.59
10.50	1.04	0.02	0.02	43.12
10.75	1.05	0.03	0.02	41.49
11.00	1.05	0.03	0.02	38.85
11.25	1.05	0.03	0.03	37.27
11.50	1.05	0.03	0.03	35.16
11.75	1.05	0.03	0.03	32.70
12.00	1.06	0.03	0.03	31.08

(*Continued*)

TABLE 12.16 (*Continued*)

Aerodynamic Coefficients of the E169 (14.4%) Airfoil

α	C_z	C_x	C_m	f
12.25	1.06	0.04	0.03	29.42
12.50	1.06	0.04	0.03	27.68
12.75	1.06	0.04	0.03	26.16
13.00	1.07	0.04	0.03	24.85
13.25	1.07	0.05	0.03	23.61
13.50	1.07	0.05	0.03	22.43
13.75	1.07	0.05	0.03	21.39
14.00	1.08	0.05	0.03	20.39
14.25	1.08	0.06	0.03	19.28
14.50	1.08	0.06	0.03	18.25
14.75	1.07	0.06	0.03	17.20
15.00	1.07	0.07	0.03	16.26
15.25	1.07	0.07	0.02	15.39
15.50	1.07	0.07	0.02	14.68
15.75	1.07	0.08	0.02	14.00
16.00	1.06	0.08	0.02	13.18
16.25	1.05	0.09	0.02	12.10
16.50	1.03	0.09	0.01	11.11
16.75	1.02	0.10	0.01	10.23
17.00	0.99	0.11	0.01	9.25
17.25	0.97	0.12	0.00	8.41
17.50	0.95	0.12	−0.01	7.60
Re = 500,000				
−16.75	−1.09710	0.08590	−0.00890	−12.8
−16.50	−1.10280	0.08157	−0.01080	−13.5
−16.25	−1.10830	0.07726	−0.01270	−14.3
−16.00	−1.11390	0.07297	−0.01450	−15.3
−15.75	−1.11830	0.06895	−0.01620	−16.2
−15.50	−1.12270	0.06500	−0.01780	−17.3
−15.25	−1.12800	0.06111	−0.01930	−18.5
−15.00	−1.13230	0.05747	−0.02050	−19.7
−14.75	−1.13600	0.05392	−0.02180	−21.1
−14.50	−1.13760	0.05067	−0.02280	−22.5
−14.25	−1.14800	0.04654	−0.02400	−24.7
−14.00	−1.15720	0.04259	−0.02520	−27.2
−13.75	−1.16390	0.03897	−0.02620	−29.9
−13.50	−1.16660	0.03587	−0.02700	−32.5
−13.25	−1.16650	0.03319	−0.02750	−35.1
−13.00	−1.16470	0.03082	−0.02770	−37.8
−12.75	−1.16070	0.02884	−0.02750	−40.2

(Continued)

TABLE 12.16 (*Continued*)

Aerodynamic Coefficients of the E169 (14.4%) Airfoil

α	C_z	C_x	C_m	f
−12.50	−1.15700	0.02703	−0.02700	−42.8
−12.25	−1.16420	0.02478	−0.02540	−47.0
−12.00	−1.16260	0.02343	−0.02330	−49.6
−11.75	−1.15750	0.02243	−0.02090	−51.6
−11.50	−1.14940	0.02165	−0.01850	−53.1
−11.25	−1.14170	0.02067	−0.01600	−55.2
−11.00	−1.13060	0.01971	−0.01410	−57.4
−10.75	−1.11450	0.01902	−0.01260	−58.6
−10.50	−1.10070	0.01810	−0.01090	−60.8
−10.25	−1.08310	0.01740	−0.00960	−62.2
−10.00	−1.06630	0.01662	−0.00820	−64.2
−9.75	−1.04860	0.01586	−0.00690	−66.1
−9.50	−1.02980	0.01517	−0.00570	−67.9
−9.25	−1.01050	0.01451	−0.00450	−69.6
−9.00	−0.99100	0.01385	−0.00330	−71.6
−8.75	−0.97090	0.01325	−0.00230	−73.3
−8.50	−0.94980	0.01273	−0.00120	−74.6
−8.25	−0.92840	0.01225	−0.00020	−75.8
−8.00	−0.90670	0.01181	0.00070	−76.8
−7.75	−0.88440	0.01143	0.00170	−77.4
−7.50	−0.86170	0.01109	0.00260	−77.7
−7.25	−0.83870	0.01079	0.00350	−77.7
−7.00	−0.81570	0.01050	0.00430	−77.7
−6.75	−0.79270	0.01023	0.00520	−77.5
−6.50	−0.76960	0.00999	0.00610	−77.0
−6.25	−0.74650	0.00978	0.00700	−76.3
−6.00	−0.70970	0.00958	0.00510	−74.1
−5.75	−0.66850	0.00938	0.00220	−71.3
−5.50	−0.62860	0.00921	−0.00020	−68.3
−5.25	−0.58280	0.00903	−0.00390	−64.5
−5.00	−0.54400	0.00891	−0.00600	−61.1
−4.75	−0.51770	0.00891	−0.00530	−58.1
−4.50	−0.49190	0.00895	−0.00460	−55.0
−4.25	−0.46630	0.00891	−0.00400	−52.3
−4.00	−0.43970	0.00890	−0.00350	−49.4
−3.75	−0.41340	0.00888	−0.00310	−46.6
−3.50	−0.38630	0.00886	−0.00270	−43.6
−3.25	−0.35930	0.00883	−0.00240	−40.7
−3.00	−0.33210	0.00881	−0.00210	−37.7
−2.75	−0.30480	0.00880	−0.00190	−34.6

(*Continued*)

TABLE 12.16 (*Continued*)

Aerodynamic Coefficients of the E169 (14.4%) Airfoil

α	C_z	C_x	C_m	f
−2.50	−0.27740	0.00877	−0.00160	−31.6
−2.25	−0.24980	0.00877	−0.00140	−28.5
−2.00	−0.22230	0.00874	−0.00120	−25.4
−1.75	−0.19450	0.00873	−0.00110	−22.3
−1.50	−0.16690	0.00870	−0.00090	−19.2
−1.25	−0.13890	0.00871	−0.00080	−15.9
−1.00	−0.11140	0.00868	−0.00060	−12.8
−0.75	−0.08350	0.00870	−0.00040	−9.6
−0.50	−0.05570	0.00867	−0.00030	−6.4
−0.25	−0.02780	0.00866	−0.00010	−3.2
0.00	0.00000	0.00869	0.00000	0.0
0.25	0.02780	0.00866	0.00010	3.2
0.50	0.05570	0.00867	0.00030	6.4
0.75	0.08350	0.00870	0.00040	9.6
1.00	0.11140	0.00868	0.00060	12.8
1.25	0.13890	0.00871	0.00080	15.9
1.50	0.16680	0.00870	0.00090	19.2
1.75	0.19450	0.00873	0.00110	22.3
2.00	0.22230	0.00874	0.00120	25.4
2.25	0.24980	0.00877	0.00140	28.5
2.50	0.27740	0.00877	0.00160	31.6
2.75	0.30480	0.00880	0.00190	34.6
3.00	0.33210	0.00881	0.00210	37.7
3.25	0.35930	0.00883	0.00240	40.7
3.50	0.38630	0.00886	0.00270	43.6
4.00	0.43970	0.00890	0.00350	49.4
4.25	0.46630	0.00891	0.00400	52.3
4.50	0.49190	0.00895	0.00460	55.0
4.75	0.51760	0.00891	0.00540	58.1
5.00	0.54400	0.00891	0.00600	61.1
5.25	0.58270	0.00902	0.00390	64.6
5.50	0.62860	0.00921	0.00020	68.3
5.75	0.66850	0.00938	−0.00220	71.3
6.00	0.70950	0.00958	−0.00500	74.1
6.25	0.74660	0.00978	−0.00710	76.3
6.50	0.76980	0.00999	−0.00620	77.1
6.75	0.79280	0.01023	−0.00530	77.5
7.00	0.81580	0.01050	−0.00440	77.7
7.25	0.83880	0.01079	−0.00350	77.7
7.50	0.86180	0.01109	−0.00260	77.7

(*Continued*)

TABLE 12.16 (*Continued*)

Aerodynamic Coefficients of the E169 (14.4%) Airfoil

α	C_z	C_x	C_m	f
7.75	0.88450	0.01143	−0.00170	77.4
8.00	0.90680	0.01181	−0.00080	76.8
8.25	0.92850	0.01225	0.00020	75.8
8.50	0.94980	0.01273	0.00120	74.6
8.75	0.97090	0.01325	0.00230	73.3
9.00	0.99110	0.01385	0.00330	71.6
9.25	1.01050	0.01450	0.00450	69.7
9.50	1.02970	0.01517	0.00570	67.9
9.75	1.04850	0.01586	0.00690	66.1
10.00	1.06610	0.01661	0.00820	64.2
10.25	1.08310	0.01740	0.00960	62.2
10.50	1.10060	0.01809	0.01090	60.8
10.75	1.11430	0.01901	0.01270	58.6
11.00	1.13050	0.01970	0.01410	57.4
11.25	1.14170	0.02065	0.01610	55.3
11.50	1.14930	0.02164	0.01850	53.1
11.75	1.15760	0.02241	0.02090	51.7
12.00	1.16290	0.02339	0.02330	49.7
12.25	1.16430	0.02476	0.02540	47.0
12.50	1.15730	0.02700	0.02700	42.9
12.75	1.16130	0.02877	0.02760	40.4
13.00	1.16420	0.03086	0.02770	37.7
13.25	1.16590	0.03325	0.02750	35.1
13.50	1.16670	0.03585	0.02700	32.5
13.75	1.16550	0.03879	0.02620	30.0
14.00	1.15820	0.04249	0.02520	27.3
14.25	1.14700	0.04667	0.02390	24.6
14.50	1.14000	0.05044	0.02280	22.6
14.75	1.13800	0.05374	0.02170	21.2
15.00	1.13350	0.05736	0.02050	19.8
15.25	1.12910	0.06104	0.01920	18.5
15.50	1.12440	0.06489	0.01770	17.3
15.75	1.11930	0.06889	0.01610	16.2
16.00	1.11420	0.07294	0.01450	15.3
16.25	1.10950	0.07716	0.01260	14.4
16.50	1.10400	0.08148	0.01070	13.5
16.75	1.09960	0.08563	0.00890	12.8

N.B.: the pitching moment is referred to $x = 0.25$; $y = 0$.

Bibliography

Abbot I., Von Doehnhoff A., *Theory of Wing Sections*, Dover Publications Inc., New York, 1958.

Althaus, D., *Profilpolaren für den Modellflug – Windkanalmessungen an Profilen im kritischen Reynoldszahlbereich*, Band 1, Neckar Verlag GmbH, Villingen, 1980.

Althaus, D., *Profilpolaren für den Modellflug – Windkanalmessungen an Profilen im kritischen Reynoldszahlbereich*, Band 2, Neckar Verlag GmbH, Villingen, 1986.

Bertin J., *Aerodynamics for Engineers*, 4 edition, Prentice Hall, Englewood Cliffs, NJ, 2002.

Le Gourière D., *L'Énergie Éolienne – Théorie, conception et calcul practique des installations*, deuxième édition, Eyrolles, Paris, 1982.

Nuovo Colombo, *Manuale dell'ingegnere*, 82esima edizione, Editoriale Hoepli, Milano, 1985.

Selig M.S., *UIUC Airfoil coordinates database*, online open access site of the University of Illinois, Collection of Profiles 1995–2014, http://aerospace.illinois.edu/m-selig/ads/coord_database.html#E.

Selig M.S., McGranahan, B.D., Wind tunnel aerodynamic tests of six airfoils for use on small wind turbines, *ASME Journal of Solar Energy Engineering* 126, 986–1001, 2004. Free PDF copy available here: http://m-selig.ae.illinois.edu/pubs/SeligMcGranahan-2004-JSEE-SixLRNAirfoils.pdf.

Index

9 781138 570191